Cyber Harassment and Policy Reform in the Digital Age:

Emerging Research and Opportunities

Ramona S. McNeal
University of Northern Iowa, USA

Susan M. Kunkle
Kent State University, USA

Mary Schmeida
Kent State University, USA

A volume in the Advances in Information Security, Privacy, and Ethics (AISPE) Book Series

IGI Global
DISSEMINATOR OF KNOWLEDGE

Published in the United States of America by
IGI Global
Information Science Reference (an imprint of IGI Global)
701 E. Chocolate Avenue
Hershey PA, USA 17033
Tel: 717-533-8845
Fax: 717-533-8661
E-mail: cust@igi-global.com
Web site: http://www.igi-global.com

Library of Congress Cataloging-in-Publication Data

Names: McNeal, Ramona S., author. | Kunkle, Susan M., 1948- author. | Schmeida, Mary, 1957- author.
Title: Cyber harassment and policy reform in the digital age : emerging research and opportunities / by Ramona S. McNeal, Susan M. Kunkle, and Mary Schmeida.
Description: Hershey : Information Science Reference, [2018]
Identifiers: LCCN 2017038368| ISBN 9781522552857 (hardcover) | ISBN 9781522552864 (ebook)
Subjects: LCSH: Cyberbullying. | Cyberstalking. | Cyberbullying--Law and legislation.
Classification: LCC HV6773.15.C92 M36 2018 | DDC 364.15/8--dc23 LC record available at https://lccn.loc.gov/2017038368

This book is published in the IGI Global book series Advances in Information Security, Privacy, and Ethics (AISPE) (ISSN: 1948-9730; eISSN: 1948-9749)

British Cataloguing in Publication Data
A Cataloguing in Publication record for this book is available from the British Library.

Advances in Information Security, Privacy, and Ethics (AISPE) Book Series

ISSN:1948-9730
EISSN:1948-9749

Editor-in-Chief: Manish Gupta, State University of New York, USA

MISSION

As digital technologies become more pervasive in everyday life and the Internet is utilized in ever increasing ways by both private and public entities, concern over digital threats becomes more prevalent.

The **Advances in Information Security, Privacy, & Ethics (AISPE) Book Series** provides cutting-edge research on the protection and misuse of information and technology across various industries and settings. Comprised of scholarly research on topics such as identity management, cryptography, system security, authentication, and data protection, this book series is ideal for reference by IT professionals, academicians, and upper-level students.

COVERAGE

- Access Control
- Cookies
- Cyberethics
- Security Information Management
- Internet Governance
- Technoethics
- Device Fingerprinting
- Global Privacy Concerns
- Risk Management
- Information Security Standards

IGI Global is currently accepting manuscripts for publication within this series. To submit a proposal for a volume in this series, please contact our Acquisition Editors at Acquisitions@igi-global.com or visit: http://www.igi-global.com/publish/.

The Advances in Information Security, Privacy, and Ethics (AISPE) Book Series (ISSN 1948-9730) is published by IGI Global, 701 E. Chocolate Avenue, Hershey, PA 17033-1240, USA, www.igi-global.com. This series is composed of titles available for purchase individually; each title is edited to be contextually exclusive from any other title within the series. For pricing and ordering information please visit http://www.igi-global.com/book-series/advances-information-security-privacy-ethics/37157. Postmaster: Send all address changes to above address.

Titles in this Series

For a list of additional titles in this series, please visit:
https://www.igi-global.com/book-series/advances-information-security-privacy-ethics/37157

Critical Research on Scalability and Security Issues in Virtual Cloud Environments
Shadi Aljawarneh (Jordan University of Science and Technology, Jordan) and Manisha Malhotra (Chandigarh Universit, India)
Information Science Reference • ©2018 • 341pp • H/C (ISBN: 9781522530299) • US $225.00

The Morality of Weapons Design and Development Emerging Research and Opportunities
John Forge (University of Sydney, Australia)
Information Science Reference • ©2018 • 216pp • H/C (ISBN: 9781522539841) • US $175.00

Advanced Cloud Computing Security Techniques and Applications
Ihssan Alkadi (Independent Researcher, USA)
Information Science Reference • ©2018 • 350pp • H/C (ISBN: 9781522525066) • US $225.00

Algorithmic Strategies for Solving Complex Problems in Cryptography
Kannan Balasubramanian (Mepco Schlenk Engineering College, India) and M. Rajakani (Mepco Schlenk Engineering College, India)
Information Science Reference • ©2018 • 302pp • H/C (ISBN: 9781522529156) • US $245.00

Information Technology Risk Management and Compliance in Modern Organizations
Manish Gupta (State University of New York, Buffalo, USA) Raj Sharman (State University of New York, Buffalo, USA) John Walp (M&T Bank Corporation, USA) and Pavankumar Mulgund (State University of New York, Buffalo, USA)
Business Science Reference • ©2018 • 360pp • H/C (ISBN: 9781522526049) • US $225.00

Detecting and Mitigating Robotic Cyber Security Risks
Raghavendra Kumar (LNCT Group of College, India) Prasant Kumar Pattnaik (KIIT University, India) and Priyanka Pandey (LNCT Group of College, India)
Information Science Reference • ©2017 • 384pp • H/C (ISBN: 9781522521549) • US $210.00

For an entire list of titles in this series, please visit:
https://www.igi-global.com/book-series/advances-information-security-privacy-ethics/37157

IGI Global
DISSEMINATOR OF KNOWLEDGE

701 East Chocolate Avenue, Hershey, PA 17033, USA
Tel: 717-533-8845 x100 • Fax: 717-533-8661
E-Mail: cust@igi-global.com • www.igi-global.com

Table of Contents

Preface

A quick scan of news headlines from the last several years finds that trending topics included stories related to Hillary Clinton's email servers, Donald Trump's Twitter use, net neutrality and recently, the #MeToo movement and the backlash against YouTube celebrity, Logan Paul, for posting a video of the body of an apparent suicide victim. These headlines illustrate how pervasive the Internet has become. The last few months have brought us other new stories in a similar vein. Among them is the story concerning Andrew Finch of Wichita, Kansas who was shot and killed after opening the door of his residence to police who were responding to a 'prank' call. The call alleged that there was a domestic dispute at his home resulting in one family member being shot and others being taken hostage. This form of 'prank" that uses the Internet or smartphones to report an emergency that sends armed police to an individual's home is known as 'swatting.' (Ellis, 2017; Buxton, 2017). Other stories include CNN host, Anderson Cooper's Twitter account being hacked, and a tweet being sent from it referring to President Trump as a 'tool' and a "loser' (Wemple, 2017) and Representative Joe Barton (R-TX) choosing not to run for reelection in 2018 after it was revealed that he had sent sexual explicit photographs, videos and messages to a woman. Even though he argued that the relationship was consensual, he announced he would be retiring after nude photos he sent her appeared online (Diaz, 2017). Additional stories include that of Mallory Grossman, a 6th grader who committed suicide in July 2017 after being bullying by her classmates, thorough text messages, Instagram and Snapchat (Stump, 2017). Each of these incidences represent a form of cybercrime ranging from swatting to revenge porn and while the consequences of these crimes vary, they are all examples of cyber aggression.

Although a concise definition is needed, most studies on cyber aggression describe it as behavior involves purposefully harming other individuals using online technology (Sontag, et al, 2011). Cyber aggression does not represent a new and emerging issue of concern. The media began covering this topic in the early 2000's when Ryan Patrick Halligan committed suicide in 2003 after being subjected to online taunts by classmates about his sexuality (NoBullying.com, 2015). His death resulted from cyberbullying, which represents an earlier form of cyber aggression. Other examples of how technology has been used to hurt others include cyberstalking and cyber-harassment. As online technology evolves, newer forms of cyber aggression including sexting, revenge porn and doxing have also developed. Acts of aggression did not begin with the invention of the Internet, traditional forms of aggression such as bullying have been described in the literature as early as the 1800s in novels such as *Oliver Twist*. While online aggression represents an evolution of traditional forms of aggression, it has not become a replacement for them. Individuals who engage in crimes of aggression are using online methods to augment tradition forms of aggression. For example, both Ryan Patrick Halligan and Mallory Grossman were victimized by their classmates in person as well as online. As a further example, consider the crime of stalking. According to the National Crime Victims Survey (Baum, et al, 2009), 14 out of every 1000 adults in the United States become a stalking victim with 26.1 percent of these stalking victims also experiencing cyberstalking or unwanted contact or monitoring through electronic devices. The Internet is not being used to replace traditional means for harming others, it is just providing stalkers and others additional avenues for victimization.

The combination of traditional and online acts of aggression has social and political ramifications. How should the government respond? Legislative action in the United States has varied. For more traditional forms of aggression and subsequently, their cyber counterparts, laws have been adopted. Currently, all 50 states and the federal government have enacted statutes aimed at protecting the victims of stalking or harassment. In addition, some states are passing new laws that include additional language for electronic communication, however others continue to use existing laws to combat violence against individuals in cyberspace (National Council of State Legislatures [NCSL], 2013). While criminal laws have been adopted for some forms of cyber aggression, legislation action is needed to address others. For example, there currently are no swatting laws at the federal level. A federal anti-swatting law, the Prioritizing Online Threat Enforcement Act of 2015, was proposed in 2015 but failed to make it out of committee (Buxton, 2017). There are a few states, such as California, which has a swatting law, but this type of law is not widespread (Ellis, 2017).

Limited legislative response in the United States to acts of cyber aggression has raised concerns among the American public. Research (DeMatteo et al, 2017) suggests that the American public believes that more can be done. This study found that respondents disagreed with current statues governing cyberstalking and believe that cyberstalking should be treated as a distinct crime from stalking. Similarly, a Pew Research Center found that the public (62% of those surveyed) felt that cyber harassment was a major problem and 43% felt law enforcement did not take it seriously enough. Although many believed that we need to do more to stop online harassment, there were conflicting beliefs on how to best precede. Among those surveyed, 64% thought that online platforms should play a major role in stopping online harassment; 60% felt that online bystanders should play a major role and 49% thought law enforcement should play a major role (Duggan, 2017). This poses an interesting question that provides the focus of this book, "if the American public believes that we need to do more to stop online aggression, why are some states taking a lead in combating online violence through new laws while others are relying on existing laws or doing nothing?"

One piece of the puzzle in explaining differences in state-level response is federalism. In the United States, laws (on the same issue) can be enacted and implemented at different levels of government. This form or government has both advantages and disadvantages. Among the advantages is that encourage experimentation with public policies by the states resulting in new policy solution and the transference of best practices (Mossberger, 2000). Federalism can also have a downside with unequal policies among the states as well as among local jurisdictions. While federalism can explain why different policy are adoption among the states for the same issue, what factors influence how lawmakers decide which policy to eventually adopted? The literature (Carmines & Stimson, 1989) on criminal justice policy argues that political factors play an important role in policy adoption of criminal statutes while the literature on policy diffusion points to a combination of political influences, state resources, and the demands or needs within the states (Mooney & Lee, 1995) with the significance of each factors varying based on the issue. To examine which of these factors take the lead in the adoption of cyber aggression laws,

Chapter 5 provides a statistical analysis of factors influencing the adoption of cyberbullying, cyber-harassment and cyberstalking laws at the state-level for 2007 through 2015. Pooled cross-sectional time series data that controls for variation between states and over time is used. This chapter also analyzes variables that influences that adoption of revenge porn laws but because these policies are relatively new, a cross-sectional analysis is presented for the year 2016.

In considering state-level online aggression policy, it is important understand the actions of all players in the federal system. Policymakers at the state-level do not act in a vacuum, their policy options are constrained by non-institutional actors such as the media, public opinion and interest groups as well as other institutional actors such as the Congress and the Supreme Court. For example, a discussion of both cyberstalking and online harassment should include a historical overview of the Violence Against Women Act (VAWA), considered a milestone in the federal legislative and legal response to gender based violence and the victimization of women and families. VAWA is recognized as having a significant affect in influencing the development of legislation and services for women and minimalized populations in all 50 states, territories, and tribal jurisdictions. In addition, a discussion of laws governing cyberbullying would be incomplete without a discussion of court decision concerning the First Amendment right to freedom of speech. The courts have, at times, sided with school districts regarding policies that constrain harmful online speech. In other instances, such as *Emmet v. Kent School District No. 415* (2000), decided by the U. S. District Court for the Western District of Washington and *Layshock v. Hermitage School District* (2006), decided by the Third Circuit Court, the rulings were in favor of students who argued their freedom of speech had been violated. Chapter 3 places the importance of other institutional actors in perspective through its discussion of federal and state-level policies as well as proving an overview of several influential court decisions regarding the First Amendment.

The history of traditional forms of aggression provides insight into why some states are laggards when it comes to the adoption of criminal laws limiting acts of online aggression. There is news coverage from the 1800s detailing accounts of suicide that resulted because of bullying. Nevertheless, bullying was treated by public officials as a 'rite of passage' for young boys (Allanson, Lester, & Notar, 2015). It took the tragic school shooting at Columbine, to spur the first school bullying policy in the United States. Similarly, it was in the 1990s that the American public first began to consider stalking a public problem and not a private matter. What is it about these two crimes that can explain the disconnect between their actual consequences and public perception? One answer could be the type of victims. Certain groups including women, youth, sexual minorities and racial/ethnic minorities are more likely to become victims of these crimes. Schneider and Ingram (1997) argue that policy design can be explain by two characteristics of the group targeted by a policy. The first is political power (strong or weak) and public perception (positive or negative). Schneider and Ingram (1997) classified women and children as *dependents*; target populations with weak political power but positive public construct. Policy aimed at dependents was argued to undersupply benefits and oversupply burdens and typically took the form of either actions directed at changing their behavior or public assistance. Classification of sexual minorities has changed over time. At one point, they would have been considered *deviants*; target populations with weak political power and negative public construct. Policy targeting deviants is argued to undersupplied benefits and oversupplied burdens and would typically take the form of punishment. The Schneider and Ingram (1997) typology, would suggest that it is not surprising that both the states and federal government were slow to adopt policy protecting victims from acts of aggression. This book does not provide a chapter that specifically examines whether characteristics or likely victims influences the type of online aggression law adopted. Nevertheless, it provides a foundation for discussing this topic. Chapter 2 examines who is more likely to be victimized through an analysis of characteristics of teens that increase their chances of becoming a victim. In addition, Chapter 6 examines whether cyber activities including posting pictures on the Internet or reviling party affiliation online are more likely to increase victimization than individual characteristics.

Finally, an emphasis of this book is on the importance of federalism in American policy adoption. One of the advantages attributed to federalism is that it allows for state-level policy experimentation and the spread of best practices. Chapter 4 discusses best practices based on what what has been learned from different policies implemented by local school boards to control cyberbullying. Additional, Chapter 6 provides an overview of the literature on what each member of the larger community can do to diminish the spread of cyber aggression. Recommendations are discussed for policymakers, law enforcements, the public and parents.

ORGANIZATION OF THE BOOK

The book is organized into six chapters. A brief description of each of the chapters follows:

Chapter 1 acts as an introduction to the concept of online (cyber) aggression. It provides the history and definition of traditional forms of aggression including bullying, stalking and harassing along with their online counterparts: cyberbullying, cyberstalking and cyber-harassment. It also discusses how the expansion of telecommunication technology has resulting in the opportunity for new forms of online aggression. Finally, this chapter provides an overview of newer forms of cyber aggression such as doxing, revenge porn, sexting and sextortion.

Chapter 2 discusses subgroups that are more likely to become victims of cyber aggression. More specifically, it asks the question, "given that research has shown that sexual minority youth are more likely to be victims of tradition forms of aggression, does the same hold true for online aggression? This chapter begins with an overview of the literature on sexual minorities and victimization. It finishes by using multivariate statistical methods and survey data from the Pew Research Center for the year 2014 to examine which demographic variables (including sexual orientation) impact the likelihood of being a victim of online aggression.

Chapter 3 illustrates the importance of federalism in explaining variation of state-level policy adoption to curtail forms of online aggression. It presents the federal and state-local legislative response to several forms of cyber aggression---stalking, harassment, and bullying. The chapter highlights federal efforts, such as the federal Violence against Women Act which have influenced state-level action on traditional and online crimes of aggression. It also discusses why certain states are acting as bellwethers on the issue of online aggression, while at the same time there are laggard states not yet entirely on board in passing legislation aligned with the advancement of technology used in cyber aggression. Finally, it explains how court cases have shaped policy decisions on the issue of cyberbullying and presents several key cases.

Chapter 4 considers the topic of best practices. An argument in favor of federalism is that it allows states or other subgovernments to experiment with policy solutions which may resulting in the adoption of better policy alternatives by other governments. It begins with a discussion of the literature on strategies being adopted at the school board level to limit the spread of cyberbullying and concludes with an overview of current evaluation research comparing recent policies being implemented by local schools.

Chapter 5 discusses the main question of this book through an examination of state-level factors suggested by the literature on policy adoption and diffusion. It examines the influence of state-level variables on policy adoption for four different forms of cyber aggression. For laws addressing cyberbullying, cyber-harassment and cyberstalking, this chapter explores the level of legislative action concerning the update and/or pass new laws for the years, 2007 through 2015. Pooled cross-sectional time series data that controls for variation between states and over time is used. Revenge porn laws are examined but because they are relatively new, a cross-sectional analysis will be presented for the year 2016.

Chapter 6 focuses on the argument that online aggression cannot be solved by anyone individual or group. It begins with an overview of the literature on what each member of the larger community can do to diminish the spread of cyber aggression. It concludes by analyzing the effectiveness of recommendations for individuals for protecting themselves from becoming a victim of online aggression as well as strategies for parents to protect their children from becoming a victim of cyberbullying. Multivariate statistical methods and survey data from the Pew Research Center for the years 2013 and 2014 were used in this analysis.

Ramona McNeal
University of Northern Iowa, USA

Susan Kunkle
Kent State University, USA

Mary Schmeida
Kent State University, USA

REFERENCES

Allanson, P., Lester, R., & Notar, C. (2015). A history of bullying. *International Journal of Education and Social Science*, *12*(2), 31–36.

Baum, K., Catalano, S., Rose, K., & Rand, M. (2009). *Stalking Victimization in the United States* (NCJ 224527). Bureau of Justice Statistics Special Report. Washington, DC: U.S. Department of Justice. Retrieved May 9, 2014, from http://www.bjs.gov/index.cfm?ty=pbdetail&iid=365

Buxton, M. (2017, March 28). *Meet the congresswomen on the Frontlines of the fight against online abuse*. Retrieved July 19, 2017, from http://www.refinery29.com/2017/03/147268/katherine-clark-online-abuse-interview

Carmines, E., & Stimson, J. (1989). *Issue evolution: Race and the transformation of American politics*. Princeton, NJ: Princeton University Press.

DeMatteo, D., Wagage, S., & Fairfax-Columbo, J. (2017). Cyberstalking: Are we on the same (web)page? A comparison of statutes, case law, and public perception. *Journal of Aggression, Conflict and Peace Research*, *9*(2), 83–94. doi:10.1108/JACPR-06-2016-0234

Diaz, K. (2017, November 30). *Congressman Joe Barton, hit with 'sexting' revelation, bows out of 2018 race*. Retrieved December 10, 2017, from http://www.chron.com/news/politics/article/Congressman-Joe-Barton-hit-with-sexting-12395264.php

Duggan, M. (2017, July 11). *Online harassment 2017*. Retrieved December 15, 2017, from http://www.pewinternet.org/2017/07/11/online-harassment-2017/

Ellis, R. (2017, December 31). *Swatting case poses legal challenges for police, prosecutors*. Retrieved January 3, 2018, from http://www.cnn.com/2017/12/31/us/swatting-legal-ramifications/index.html

Emmet v. Kent School District No. 415 92 F. Supp. 2d 1088 (W.D. Wash. 2000).

Layshock v. Hermitage School District 650 F.3d 205 (3rd Cir. 2006).

Mooney, C., & Lee, M. (1995). Legislating morality in the American states: The case of pre-Roe abortion regulation reform. *American Journal of Political Science*, *39*(3), 599–627. doi:10.2307/2111646

Mossberger, K. (2000). *The politics of ideas and the spread of enterprise zones*. Washington, DC: Georgetown University Press.

National Council of State Legislatures. (2013). *Cyberstalking and Cyber harassment Laws*. Retrieved May 7, 2014, from http://www.ncsl.org/research/telecommunications-and-information-technology/cyberstalking-and-cyberharassment-laws.aspx

NoBullying.com. (2015). *Ryan Halligan loses his life to taunts, rumors and cyber bullying*. Retrieved July 11, 2015, from https://nobullying.com/ryan-halligan

Schneider, A., & Ingram, H. (1997). *Policy design for democracy*. Lawrence, KS: University Press of Kansas.

Sontag, L., Clemans, K., Graber, J., & Lyndon, S. (2011). Traditional and cyber aggressors and Victims: A comparison of psychosocial characteristics. *Journal of Youth and Adolescence*, *40*(4), 392–404. doi:10.100710964-010-9575-9 PMID:20680425

Stump, S. (2017, August 2). *Parents of bullied 12-year-old girl who committed suicide: 'We want to honor her.'* Retrieved August 2, 2017, from http://www.msn.com/en-us/health/healthtrending/parents-of-bullied-12-year-old-girl-who-committed-suicide-we-want-to-honor-her/arAApiPSl?li=BBnba9O&ocid=mailsignout

Wemple, E. (2017, December 13). *Anderson Cooper's boring Twitter account suddenly perks up. CNN claims hacking*. Retrieved December 20, 2017, from https://www.washingtonpost.com/blogs/erik-wemple/wp/2017/12/13/anderson-coopers-boring-twitter-account-suddenly-perks-up-cnn-claims-hacking/?utm_term=.6b53c08df6ce

Chapter 1
Defining Online Aggression:
From Cyberbullying to Nonconsensual Pornography

ABSTRACT

The internet has become an inescapable part of our lives, and while it makes our lives easier, it also exposes us to online threats ranging from identity theft to denial of service to phony lottery/sweepstake scams. Among these online threats are those that are carried out with the direct intent of harming another person or group of individuals. This category of crimes is referred to as cyber aggression and includes cyberbullying, cyber-harassment, and cyberstalking. As technology expands, so does the opportunity for new forms of online aggression such as doxing and revenge porn. It is becoming difficult to keep up with new trends in acts of online aggression or distinguish between cybercrimes that appear to have similar definitions. This chapter acts as an introduction to online aggression by providing an overview of older and emerging forms of cyber aggression.

INTRODUCTION

Acts of cyber aggression, like other Internet crimes, have unique features making them more difficult to address. First, the Internet empowers perpetrators of cybercrimes by helping to mask their identity. Individuals who wish to remain anonymous can create multiple fake email accounts or use messaging

DOI: 10.4018/978-1-5225-5285-7.ch001

accounts such as Whisper, all of which conceal identity and make it more difficult for law enforcement to identify suspects and gather evidence. Even if law enforcement can determine the Internet protocol (IP) address of a computer from which a crime was committed, it can be difficult, if not impossible, to establish who used it (Jany, 2016). In addition, anonymity can increase the likelihood of someone committing a cybercrime by reducing inhibitions and individual restraint (D'Ovidio & Doyle, 2003). Second, since cybercrimes are not constrained by geographical boundaries, it becomes challenging to determine which level of government has jurisdiction. One example is the story of Canadian teen, Amanda Todd, who committed suicide in 2012. While in the 8th grade, a cybercrime perpetrator posted explicit pictures of Amanda online and sent them to her classmates. Although Amanda's family transferred her to different schools following the initial and subsequent incidents, the perpetrator continued to distribute pictures to classmates at each new school. After two years of being taunted and bullied by other students, Amanda committed suicide (Grenoble, 2012; Nobullying.com, 2017). Her tormentor, Aydin Coban, a 35-year-old Dutch male, was identified in 2014. Coban was found to have several victims in the United States., United Kingdom, Canada and the Netherlands. His modus operandi was to manipulate under-aged girls to pose seductively in front of a webcam and then use the images to extort, taunt and bully his victims (Nobullying.com, 2017).

Though Coban was arrested and sentenced in the Netherlands, jurisdictional issues can make obtaining justice more difficult. For example, who has jurisdiction when the victim lives in one jurisdiction and the offender lives in another? The answer is not always clear. The jurisdiction where a crime is tried can significantly influence legal relief for the victim. For example, the definition of a crime can and does vary among jurisdictions and in some instances what is a crime in one jurisdiction may not be a crime in another. Finally, if the cybercrime originates from a country where there is no extradition agreement with the country where the victim resides, there may be no legal resolution.

The differences in how cybercrimes are legally defined do not only vary between countries. In the United States the judicial response to cybercrimes has varied based on the level of government and the form of aggression (whether physical or cyber). For example, the federal government, all 50 states, the District of Columbia, and U.S. territories have enacted criminal laws to address stalking. However, the legal definition of stalking varies across jurisdictions. The variance in state laws concerns the element of fear and emotional distress experienced by the victim, and the intent of the stalker. States vary on the level of fear required. For example, some state laws require prosecutors to establish fear of death or serious bodily harm, while others require that prosecutors establish that the victim suffered emotional distress. Since the United States has a federal system, once a crime crosses state boundary it becomes a federal crime. In the case of interstate stalking, the crime is defined by federal law 18 U.S.C. § 2261A (Catalano 2012; Baum, Catalano, Rose, & Rand, 2009).

As with stalking, the states vary in their legal definition of other forms of aggression such as harassment and bullying. While some states are passing new laws that include additional language for electronic communication, others continue to rely on existing laws. Although older laws can be used to combat violence against individuals in cyberspace, the new laws may make it easier for state and local governments to prosecute forms of cyber violence (National Council of State Legislatures [NCSL], 2015). This raises some important questions such as, "why are some states taking a lead in combating online violence through new laws while others are relying on existing laws?" Other questions include "how do individuals best protect themselves from becoming victims of cyber aggression?" and "how do the states respond to cyber aggression?" To examine these questions, this book will explore the intersection of technology and crime, and identify the federal and states' legislative and judicial response to the evolution of cyber aggression crimes.

Chapter 1 begins with an overview of cyberbullying and merges it with the history and scope of traditional (physical) forms of bullying. There are definitional issues between bullying and cyberbullying with some researchers viewing cyberbullying as a specific type of bullying – the technology merely being the tool used to bully (U.S. Department of Health & Human Services, 2015). These issues are further discussed in this chapter. Bullying, which is a type of violence, is a pattern of behavior rather than an isolated event, and it has an adverse impact on the victim, the bully and bystanders (UNESC, 2017, p.8). It includes actions such as making threats, spreading rumors, attacking someone physically or verbally, and excluding someone from a group on purpose. According to U.S. Department of Health & Human Services (2015), cyberbullying is a specific form of bullying that takes place online.

HISTORY: BULLYING AND CYBERBULLYING

Bullying is neither a contemporary phenomenon nor a behavior unique to the United States. As a literary concept, bullying came to prominence in 1838 in Charles Dicken's *Oliver Twist* followed by Thomas Hughes' *Tom Brown's School Days* in 1857 (Srabstein & Merrick, 2013). *Tom Brown's School Days* is a story set in a public school for boys of the 1830s in which Tom and peers are tormented by a bully. According to the narrative, while the consequences of bullying were recognized as potentially debilitating, the behavior was interpreted as a rite of passage and viewed as appropriate conduct for young boys. At the same time these early literary works were portraying bullying as a 'rite of passage', the press was providing examples of its potential consequences. In 1885, at King's College in London, a young boy died because of bullying by a group of older peers. Even though the school council was pressured to investigate the incident by former students and the public, the council refused based on the belief that bullying was a normal part of a young boy's life (Allanson, Lester, & Notar, 2015). Among the earliest reports of bullying was an article in *The Times of London* August of 1862 (Koo, 2007). According to the *Times*, a sailor who had been the victim of extended, malicious and systematic bullying, took the life of the person who was responsible for the bullying (Koo, 2007).

The earliest known scientific study of bullying was published in 1897 by Frederic Burk, a professor at Clark University (National Academies of Sciences, Engineering, and Medicine, 2016). Burk, using survey data, attempted to describe, classify and explain the acts of teasing and bullying (Burk, 1897). Burk's definition of teasing most closely resembles the contemporary definition of verbal (teasing, taunting, and name-calling) and social bullying (spreading rumors, leaving someone out on purpose); his definition of bullying is analogous to physical bullying (hitting, kicking, and pinching). In the early 1970s, Professor Dan Olweus of the University of Bergen, Norway began the first large scale research project on bullying, victimization, and violence. In 1983, three adolescent boys in Norway, each with a history of victimization resulting from bullying, committed suicide. As a response to the deaths of these young boys, the Norwegian Ministry of Education initiated a public campaign to combat serious bullying in schools and classrooms. The Olweus Bullying Prevention Program (OBPP) was developed amid these tragic events (Limber, 2011).

A GLOBAL PROBLEM

In 2016, the United Nations Educational, Scientific and Cultural Organization (UNESCO) conducted a survey on school violence and bullying that spanned 18 countries. The number of respondents was 100,000 young adults; two-thirds or 76% of the sample reported that they had been victimized by bullying. It is estimated that globally 246 million children and adolescents experience school violence and bullying in some form every year. According to UNESCO, bullying harms the health and emotional well-being of children and adolescents (2017). Some of the reported physical effects of bullying include stomach pains and headaches and difficulty eating and sleeping. Youth who are bullied are also more likely than those who are not bullied to experience interpersonal difficulties, to be depressed, lonely or anxious, to have low self-esteem and to have suicidal thoughts or to attempt suicide (UNESCO, 2017).

The Global School-based Student Health Survey (GSHS) was developed by the World Health Organization (WHO) and the Centers for Disease Control and Prevention (CDC) in collaboration with UNICEF, UNESCO, and UNAIDS. GSHS is a school-based survey conducted primarily among students aged 13–17 years. The Health Behavior in School Age Children (HBSC) research network is an international alliance of researchers that collaborate on the cross-national survey of school students. The HBSC collects data every four years on 11-, 13- and 15-year-old boys' and girls' health and well-being, social environments and health behaviors. According to the data collected from GSHS and HBSC, bullying is a worldwide problem that exists in every country. Data collected from 106 countries involving youth age 13 to 15 years, and who identify with recent incident of bullying range from 7% in Tajikistan to 74% in Samoa. Additionally, 31% of teens in Europe and North America admitted to having bullied others, with a prevalence rate of one in seven (14%) in the Czech Republic and Sweden to nearly 6 in 10 (59%) in Latvia and Romania (United Nations Children's Fund, 2014). In the United States, for the 2014-2015 school years, approximately 21% of school children age12 through 18 years reported being bullied (Musu-Gillette, et al., 2017).

DEFINITIONS

Bullying is an intentional negative behavior that typically occurs with some repetitiveness and is directed against a person who has difficulty defending them self (Olweus, 2011). According to UNESCO (2017), bullying is a type of violence, a pattern of behavior rather than an isolated event, having an adverse impact on the victim, the bully and bystanders. Bullying has been defined as 'unwanted, aggressive behavior among school-aged children that involves a real or perceived imbalance of power". The behavior is repeated, or has the potential to be repeated, over time. Eisenberg and Aalsma (2005) in their review of the scientific research, identified the terms of "bullying," "harassment," and "peer victimization" as synonymous descriptions of behavior that was defined as bullying. While bullying is a type of violence, not all descriptions of aggressive behavior specifically define bullying (UNESCO, 2017). For example, peer victimization, when reflective of a fight, argument, or disagreement and devoid of a power imbalance is not bullying (U.S. Dept. of Health & Human Services, 2017b).

In 2010, the U.S. Department of Education's Office for Civil Rights (OCR) issued a letter to school districts that clarified the relationship between bullying and discriminatory harassment as found under Title VI of the Civil Rights Act of 1964; Title IX of the Education Amendments of 1972; and Title II of the Americans with Disabilities Act of 1990. As stated in the Dear Colleague letter, bullying based on race, color, national origin, sex, or disability can be a civil rights violation if it is "sufficiently serious" to constitute a hostile environment (U.S. Department of Education, 2010). The Dear Colleague directive from OCR identified harassment as verbal acts and name-calling, graphic and written statements (with or without the use of cell phones and/or the Internet), and conduct that may be physically threatening, harmful, or humiliating. Furthermore, harassment does not have to include intent to harm, to be target specific, or involve repetitive incidents. A hostile environment is created when the conduct is sufficiently severe, pervasive, or persistent to interfere with or limit a student's ability to participate in or benefit from the services, activities, or opportunities offered by a school. The intent of the letter was to remind school districts that student misconduct that falls under a school's anti-bullying policy may also trigger responsibilities under one or more of the federal antidiscrimination laws enforced by the Department's Office for Civil Rights (U.S. Department of Education, 2010).

Bullying and harassment are similar in that they involve actions that are predicated on the issues of power and control, consist of behaviors that are intended to physically or emotionally harm another person, and reflect a disparity of power between the perpetrator and victim (PACER, 2017; U.S. Department of Education, 2010). The distinction becomes when the aggressive behavior is directed at a member of a protected class (race, color, national origin, sex, or disability) the behavior is defined as (discriminatory) harassment (PACER, 2017). According to the National Center Against Bullying, bullying is a relationship problem occurring when an individual or group of people with more power, repeatedly and intentionally cause hurt or harm to another person or group of people who feel helpless to respond. Bullying is not a single episode of social rejection or dislike, a single episode of nastiness or spite, random acts of aggression or intimidation, or mutual arguments, disagreements, or fights (2017). Bullying, as broadly defined in the literature, is an aggressive act that reflects an imbalance of power between the perpetrator and victim, occurs frequently and can manifest verbally (threats or insults), with physical force (assaults), indirectly or relational (exclusion or rumormongering) and cyber (Internet and technology based) (UNESCO, 2017; Srabstein & Merrick, 2013; Olweus, 2011; Wang et.al., 2009).

UNIFORM BULLYING DEFINITION PROJECT

In 2008, the Federal Partners in Bullying Prevention Steering Committee was founded to provide federal guidance on bullying. In 2010, a summit was held which brought together various corporate leaders, academics, researchers, practitioners, parents, and youth to identify challenges to bullying prevention efforts. A critical issue emerging from the Summit concerned the variations surrounding the definition of bullying behaviors and the need for a definition that was consistent and uniform (Gladden, et.al, 2014). Subsequently, in a publication of the National Center for Injury Prevention and Control of the Centers for Disease Control and Prevention, concern was expressed that uneven definitions of bullying and varied measurement strategies have made it problematic to compare the prevalence of bullying with outcomes across studies (Hamburger, Basile, & Vivolo, 2011; Vivolo, Holt, & Massetti, 2011).

The definition of bullying developed by Gladden et al. (2014) was theoretically based on the definition developed by Olweus (1993) while responding to concerns expressed by researchers that bullying can also be a single act of aggression (Finkelhor, Turner, & Hamby, 2012). More specifically, bullying is any unwanted aggressive behavior(s) by another youth or group of youths who are not siblings or current dating partners that involves an observed or perceived power imbalance and is repeated multiple times or is highly likely to be repeated. Bullying may inflict harm or distress on the targeted youth including physical, psychological, social, or educational harm (Gladden, et.al, 2014; Institute of Medicine [IOM] & National Research Council [NRC], 2014). The key elements of this definition are: Unwanted which refers to the fact that the targeted youth wants the aggressive behaviors to stop. Aggressive behavior which is the intentional use of harmful behavior(s), threatened or actual, against another youth. Has occurred multiple times or is highly likely to be repeated means that the youth experiences multiple incidents of aggression perpetrated by a single youth or group of youths over a specified time or there is strong concern a single aggressive behavior by a youth or group of youths has a high likelihood of being followed by more incidents of aggression. A power imbalance is the attempt by the perpetrator(s) to use observed or perceived personal or situational characteristics to exert control over the targeted youth's behavior or limit the victim's ability to respond or stop their aggression. Harm is the negative experiences or injuries and can include physical cuts, bruises or pain; psychological consequences such as feelings of distress, depression or anxiety; social damage to reputation or relationship, and/or interference with education because of increased absenteeism, dropping out of school, having difficulty concentrating in class, and poor academic performance (Gladden, et. al., 2014; IOM & NRC, 2014).

MODES OF BULLYING

Bullying can be overt and direct or covert and indirect (PACER, 2017; Gladden, et. al., 2014; Institute of Medicine & National Research Council, 2014). Overt or direct bullying is when the aggressive/physical behaviors occur in the presence of the victim. Examples of overt or direct bullying include hitting or pushing the victim, and other forms of face to face interaction such as taunting or teasing. Covert or indirect bullying occurs when the aggressive behaviors are not directly communicated to the victim and include spreading false and harmful rumors, excluding the youth from events, or victimizing the youth via the Internet through smart phones, texts, and images (PACER, 2017; Gladden, et.al., 2014).

In addition to mode, bullying can be further classified based on form. There are three types of bullying (U.S. Department of Health & Human Services, 2017a; PACER, 2017). Physical bullying involves the use of physical force by the perpetrator against the victim, for example hitting, kicking, punching, spitting, tripping, pushing, and destroying someone's personal property (U.S. Department of Health & Human Services, 2017b; Gladden, et. al, 2014). Verbal bullying is oral or written communications against a target youth that is intended to cause harm, such as name-calling, taunting, verbal threats, and inappropriate sexual comments. Relational or social bullying is intended to harm the reputation and relationships of the targeted youth (U.S. Department of Health & Human Services, 2017b; PACER, 2017; Gladden, et.al, 2014). A goal of relational aggression is to isolate the youth and keep the victim from interacting with peers, for example by spreading false and harmful rumors, posting and sending inappropriate and embarrassing images of the targeted youth without permission.

WHERE DOES IT OCCUR?

Historically, bullying has occurred at school or anywhere that children played or congregated and can occur inside and outside of the classroom and on the way to and from school (UNESCO, 2017, National Academies of Sciences, Engineering, and Medicine, 2016). In school, bullying often occurs in places such as toilets, changing rooms, corridors and playgrounds where children and adolescents are less easily seen or supervised by teachers and other staff (UNESCO, 2017). Bullying, a type of violence, interferes with a youth's emotional, physical, and social welfare in schools, afterschool programs, neighborhoods, and communities (Gladden, et al, 2014). It occurs because of many different factors such as the personal characteristics of the youth, interpersonal relationships with peers and adults, and school and community norms (Gladden, et.al, 2014; IOM & NRC, 2014).

According to Cohn and Caumont (2016), American communities are more racially and ethnically diverse than in the past, and the U.S. is projected to be even more diverse in the coming decades. Concurrent with the demographic changes in American communities, are major changes in the ethnic and racial composition of schools across the country. More recently, the physical setting in which bullying is occurring has been augmented with a virtual setting facilitated through the World Wide Web. Continual advances in technology and the ever-popular social messaging, has created cyberbullying, a new type of digital electronic aggression, which takes place through chat rooms, instant messaging, social media, and other forms of digital electronic communication. Thus, any attempt to effectively respond to the bullying problem in the United States must recognize that peer group composition, shifting demographics, changing societal norms, and modern technology are all important variables that must be taken into consideration (National Academies of Sciences, Engineering, and Medicine, 2016).

Cyberbullying is defined as willful and repeated harm inflicted using electronic technology such as computers, tablets, cell phones as well as communication tools such as social media sites, chat rooms, and websites (U.S. Department of Health and Human Services, 2015). Cyber aggression, according to the Center for Disease Control, has become an emergent concern and includes sending threatening texts, posting or distributing defamatory or harassing messages, and uploading or distributing hateful or demeaning images or videos intended to harm another. Cyberbullying differs from bullying in that it can occur all day, every day and can happen any hour of the day or night. Cyberbullying messages can be posted anonymously and transmitted instantaneously to a national and international audience. The ability to delete harassing and victimizing messages, posts, pictures, and videos and the ability to trace the source of the postings is extremely difficult (U.S. Department of Health and Human Services, 2015). In addition, the likelihood of being a victim of cyberbullying is increasing as the Internet has become a part of our daily lives. A 2014 study by the Pew Research Center found about 65% of young adults between the ages of 18 and 29, who use the Internet have been subject to some degree of on-line harassment. The proportion rises to 70% for those between 18-24 years of age (Pew Research Center, 2014, p. 3).

Although the term bullying has been part of the public vocabulary since the 1800's, the term cyberbullying did not become lexicon until the death of Ryan Halligan in 2003. The 13-year-old from Vermont, hanged himself after he was taunted both online and in person regarding his sexuality. In addition, a popular girl at school pretended to like him and then posted their Internet exchanges online to humiliate him. His death resulted in the passage of early state laws meant to curb cyberbullying policy including the 2004 Vermont Act 117; 16 V.S.A. § 11(a) (32) (Nobullying.com, 2015a; Capretto, 2016; Cyberbullying Research Center, 2016). Although public concern regarding cyberbullying would dissipate, several high-profile suicides brought this cybercrime back to the forefront. In 2006, Megan Meier hanged herself a few weeks before her 14th birthday. She had been the victim of cyberbullying through her MySpace account by one of her "friends" and her friend's mother. The involvement of an adult in these online attacks of a thirteen-year-old helped to draw public attention and media coverage (Frankel, 2012). The suicide of 15-year-old Phoebe Price in 2010 brought both national and international attention to bullying in the United States. Although she lived in Massachusetts at the time of her death, she and her family had moved to the United States from Ireland. She was subjected to both intense bullying in-person and online from classmates at her new school. Following her suicide, six students were charged with a variety of crimes including statutory rape, harassment, and stalking. Since her tormentors were minors they only received community service and a suspended sentence despite the nature of their crimes. Most of the bullying took place on school grounds and many teachers and others knew about it but did nothing. Massachusetts did not have an anti-bullying law at that time and school personnel were not required to intervene (Eckholm & Zezima, 2010; Leefeldt, 2016). After her death, the state of Massachusetts enacted Chapter 92 An Act Relative To Bullying In Schools which addressed both bullying and cyberbullying (Eckholm & Zezima, 2010; Cyberbullying Research Center, 2016a).

STALKING, CYBERSTALKING, AND CYBER-HARASSMENT

According to the Stalking Resource Center, stalking is a pattern of repeated and unwanted attention, harassment, contact, or any other course of conduct directed at a specific person that would cause a reasonable person to feel fear. Stalking behaviors may include: sending repeated, intrusive, unwanted, and frightening communication to the victim by phone, mail, and/or email; repeatedly sending or leaving unwanted items, gifts, or flowers, making direct or indirect threats to do harm to the victim, the victims' relatives, friends, or pets; damaging or threatening to damage the victim's property; harassing the victim through the Internet; posting information or pictures on the internet; spreading rumors about the victim on the Internet, in a public place, or by word of mouth; and, following the victim or hiring someone to follow the victim. In other words, stalking behaviors can take on many forms, but the common denominator is that the behavior is unwanted, repetitious and often obsessive, and frequently illegal (National White Collar Crime Center, 2013). Within the literature there is consensus that research on the crimes of stalking, cyberstalking and cyber-harassment is lacking (Catalano, 2012). Moreover, crimes of this type tend to be underreported which further obscures our ability to understand the prevalence of this type of behavior (National White Collar Crime Center, 2013). A common reason for not reporting a crime of this type is because the victim classifies the action(s) as a private and/or family matter. Additionally, the victim may not recognize the behavior as criminal (Schnell & Garcia, 2013). Actions such as gift giving, sending letters or email, rumormongering, and appearing at a site where the victim is present are not, in and of themselves, are not necessarily criminal. Legislatively, it is the repetitiveness of the acts; a course of conduct that causes the victim to be fearful for personal or family safety that classifies many of the behaviors associated with stalking as illegal (Schnell & Garcia, 2013; Baum, et al, 2009).

Like bullying, stalking has had a long literary history but a limited legislative history. California became the first state to pass a stalking law in 1990 following the death of actress Rebecca Schaeffer known for her role in the television series, *My Sister Sam*. She had been stalked and killed by an obsessed fan. Other states were quick to follow with their own stalking laws (Mullen & Pathe, 2002). Schaeffer is not the only personality to be stalked and killed by an obsessed fan. Most recently, singer Christina Grimmie was shot by a stalker at a June 2016 concert in Florida (Fisher, 2017). Although media coverage of the murder of Christina Grimmie and other celebrities such as John Lennon may give the impression that stalking is primarily a danger for those in the public eye, it is a problem faced more by those in the public. Celebrity stalking is only one form of this crime. Typically, a stalking victim is female, who knows her stalker and often has had a previous romantic relationship with this individual (Spitzberg & Cupach, 2007).

The growth of the Internet, a proliferation of social media websites, and the explosion of information available online facilitates the ability of those who engage in cyberstalking behaviors to search and intimidate their victims. The expanded World Wide Web and the continual and rapid development of technology that allows individuals to access the Internet and social media accounts from any location can and does enable the constant bullying and harassment of victims (National White Collar Crime Center, 2013). The Center for Disease Control defines electronic aggression as any type of harassment or bullying (teasing, telling lies, making fun of someone, making rude or mean comments, spreading rumors, or making threatening or aggressive comments) that occurs through email, a chat room, instant messaging, a website (including blogs), or text messaging (Hertz & David-Ferdon, 2008). In 2000, 6% of Internet users ages 10-17 identified as victims of on-line harassment, defined as threats or other offensive behavior, sent on-line to someone or posted on-line. In 2005, this percentage had increased by 50%, to 9% (Wolak, Mitchell, & Finklehor, 2007a). A January 2017 survey found that among U.S adults, 41% had experienced some form of online aggression. The majority had been subject to less severe forms of harassment including name calling (27%) and being purposively embarrassed (22%). Among the more severe forms of online aggression experienced were physical threats (10%), stalking (7%), sustained harassment (7%) and sexual harassment (6%) (Pew Research Center, 2017).

Cyberstalking, more specifically, is defined as the repeated use of the Internet, e-mail, or related digital electronic communication devices and technology to annoy, alarm, or threaten a specific individual or group of individuals. Cyber-harassment differs from cyber stalking in that there is the lack of an immediate credible threat. Cyber-harassment most often pertains to threatening or harassing email messages, instant messages, or to blog entries or websites dedicated solely to tormenting an individual (National Council of State Legislatures, 2015). The terms cyber- harassment and cyberbullying are often confused and have been treated the same in the literature. Wolak et al (2007b) argue that bullying is more harmful than harassment. In addition, actions that begin as harassment can escalate to bullying. Within the literature there is also some disagreement on the definition of cyber stalking (Southworth, Finn, Dawson, Fraser & Tucker, 2007; Spence- Diehl, 2003). Spence-Diehl contends that cyber stalking is merely the tools that are used by the perpetrator to stalk – in other words the method of stalking (Spence-Diehl, 2003). The Stalking Resource Center (2003) suggests that the word cyber (or Internet) adds an element of confusion in the investigation and prosecution of crimes classified as cyber stalking and, like Spence-Diehl, prefers a definition that will address all the types of technology that are used to stalk and frighten victims. This can be captured by including in the legislation – "the use of technology to stalk' – which then will broaden the scope of the criminal behavior to include cases such as: stalkers who misuse technology to send multiple emails or text messages; perpetrators who monitor a victim's computer activity through Spyware; stalkers who track the location of a victim through a Global Positioning System (GPS) device; or stalkers who plant and make use of a hidden camera to monitor the activities and routines of their victim (Stalking Resource Center, 2017).

No matter the definition of stalking and its cyber companions, the behavior associated with the crime consists of the following elements. It is a course of conduct that means the action is repetitious and has occurred on more than one occasion. The course of conduct results in unwanted, unsolicited, unwarranted, and frightening attention and interaction for someone (Stalking Resource Center, 2017). Stalking frequently involves the misuse of multiple technologies to engage in a pattern of harassment and to facilitate the uninvited attention and contact (National Network to End Domestic Violence, Safety Net Project, 2009). These technologies can include computer and Internet technology, telephone technologies, and GPS and Location Services (Logan, 2010). Finally, the actions and behaviors that comprise the course of conduct are intended to make a reasonable person feel fear or concern about personal safety (Southworth, Finn, Dawson, Fraser & Tucker, 2007).

Forms of online aggression including cyberstalking and cyber-harassment can result in serious consequences for the victims. In extreme circumstances such as Ryan Halligan, Megan Meier and Phoebe Price, the victim committed suicide. Youth who are victims of Internet harassment are significantly more likely than those who have not been victimized to use alcohol and other drugs, receive school detention or suspension, skip school, or experience in-person victimization (Ybarra, Diener-West & Leaf, 2007). It is not only the victims who are being impacted. A recent Pew survey found that among U.S. adults 18 and older, 27% report deciding not to post something online after observing others being harassed for their posts. In addition, 13% reported cancelling an online service after witnessing others being harassed (Pew Research Center, 2017). Not only are victims being harmed emotionally, psychologically, and by damaged reputation, but their free speech is being challenged. Some individuals are becoming frightened to express their viewpoints online.

Some forms of stalking/harassment, because of their unique characteristics, have been single out as a separate form of crime. One example is 'swatting', a cybercrime that involves terrorizing another person through making a false report to police. This 'prank" involves reporting an emergency that results in armed police being sent to an unsuspecting individual's home (Buxton, 2017). This form of online crime has typically been perpetrated against celebrities including Tom Cruise, Rihanna and Miley Cyrus, but it has also been used to victimize ordinary citizens. While this crime can result in unnerving the unsuspecting victim and wasting law enforcement resources, it can also produce tragic events. One example is the unintended death of Andrew Finch who was shot and killed after opening the door of his residence to police who were responding to a 'swatting' prank that reported a domestic dispute at his home (Ellis, 2017).

REVENGE PORN AND SEXTING

The ongoing fight between reality stars, Blac Chyna and Rob Kardashian has focused recent attention to the topic of revenge porn. On July 10, 2017, Chyna was granted a temporary restraining order against her ex fiancé after he posted sexually explicit photos of her on his Instagram account and accused her of cheating on him (France, 2017). A question raised by several media outlets was whether Rob Kardashian had violated California's revenge porn law by posting these images of his former fiancée? Furthermore, if so, what were the possible and probable legal consequences?

What is revenge porn? While there are varying definitions of revenge porn or nonconsensual pornography (NCP), revenge porn is commonly defined as the distribution of sexually explicit images of an individual online without their consent. In addition to the images, identifying information about the victim such as name, phone and address are often included to encourage others to harass the victim (Citron & Franks, 2014; Lonardo, Martland, & White, 2016). The Cyber Civil Rights Initiative (2017) suggests that the phrase 'nonconsensual pornography' is more appropriate as well as the following definition, "the distribution of sexually graphic images of an individual without their consent." The preference for this terminology is because revenge is not always the motive for distributing sexually explicit images. The phrase 'revenge porn' has been used to describe a variety of crimes that have different underlying motives. The case of Amanda Todd would appear to be an example of revenge porn but is classified as an example of Sexual Extortion of Children in Cyberspace (SECC). This cybercrime is defined as, "the blackmailing of a person with the help of self-generated images of that person to extort sexual favors, money or other benefits from her/him under the threat of sharing the material beyond the consent of the depicted person" (Interagency Working Group on Sexual Exploitation of Children, 2016, p. 52). Other cybercrimes that could be confused with revenge porn include doxxing and sextortion. Doxxing or doxing involves posting personal information about another in an act of revenge. Unlike NCP, the information need not include explicit images or be sexual in nature. Sextortion, on the other hand, is a form of extortion where a threat is made to post sexual images online unless the victim agrees to take part in certain sexual activities (Buxton, 2017).

Revenge porn is often linked to sexting which involves taking a naked/partially naked picture of oneself and sending the images to another person. Typically, the transmission of these images occurs through text messaging.

If the images are hacked or circulated to others, sexting can easily turn into cyberbullying and nonconsensual pornography. This happened to Jessica Logan, an 18-year-old from Cincinnati, Ohio who committed suicide in 2008 after her former boyfriend shared nude photos that she had sent him, with other classmates. Jessica was harassed constantly by other girls in her school after her pictures were circulated (NoBullying.com, 2015b; Powell, 2016). Following her death, in 2012, the state of Ohio passed H.B.116, the Jessica Logan Act. Ohio H.B. 116 is an anti-bullying law that requires schools to have policies for both traditional and cyberbullying (Cyberbullying Research Center, 2016). It is important to note that H.B. 116 was not a sexting law. Many state legislatures and courts continue to be challenged on how to legislatively and judicially address sexting. In many states, child pornography laws, initially written for adults, are being used to prosecute teenagers that take part in sexting (Powell, 2016).

The current legal state of sexting illustrates an important problem with nonconsensual pornography; compared with other forms of cyber aggression, states have been slower to pass legislation addressing these crimes. Initially, NCP cases were often treated as copyright violations when they went to trial. Consequently, if the victim did not own a copyright on the explicit images posted, law enforcement was limited in what it could do (Hern, 2016). Additionally, while some revenge porn sites were shut down and their owners arrested, charges, other than those related to NCP, were used. One of the largest sites to be shuttered was *Is Anyone Up?* operated by Hunter Moore. Moore's site published explicit pictures of ordinary citizens and recognized celebrities and was known for including the contact information of the victims next to the pictures. Ultimately, he was arrested under federal law for paying to have phones and computers hacked to obtain photos. His site, started in 2010, was closed in 2012. Moore accepted a plea deal in 2015 where he received a two-and-a-half-year prison sentence, followed by three years of supervised release and a $2,000 fine (Ohlheiser, 2015). Moore's high-profile case helped to spur legislation across the United States. In 2013 California passed a revenge porn law with several states following suit. Nevertheless, it is still difficult to close revenge porn sites because of section 230 of the Communication Decency Act of 1996 which grants immunity to owners of websites for content that was created by others (Electronic Frontier Foundation, 2017).

Table 1 provides a summation of the forms of aggression discussed in this chapter.

Table 1. Forms of aggression

Form of Aggression	Definition
Traditional Aggression	
Bullying	An intentional negative behavior that typically occurs with some repetitiveness and is directed against a person who has difficulty defending himself or herself.
Harassment	The definition is like bullying but is classified as harassment if the negative behavior is directed at a member of a protected class (race, color, national origin, sex, or disability)
Stalking	Stalking is a pattern of repeated and unwanted attention, harassment, contact, or any other course of conduct directed at a specific person that would cause a reasonable person to feel fear.
Nonconsensual Pornography	The distribution of sexually graphic images of an individual without their consent.
Cyber Aggression	
Cyberbullying	Bullying (teasing, telling lies, making fun of someone, making rude or mean comments, spreading rumors, or making threatening or aggressive comments) using online technology.
Cyber-Harassment	Threatening or harassing email messages, instant messages, or to blog entries or websites dedicated solely to tormenting an individual.
Cyberstalking	The repeated use of the Internet, e-mail, or related digital electronic communication devices and technology to annoy, alarm, or threaten a specific individual or group of individuals.
Nonconsensual Pornography(cyber)	The distribution of sexually explicit images of an individual online without their consent.
Sexual Extortion of Children in Cyberspace (SECC)	The blackmailing of a person with the help of self-generated images of that person to extort sexual favors, money or other benefits from her/him under the threat of sharing the material with others.
Sextortion	A form of extortion where a threat is made to post sexual images online unless the victim agrees to take part in certain sexual activities.
Doxing	The posting of personal information about another in an act of revenge.
Sexting	Sending a naked/ partially naked picture of oneself to another person. Typically, this would be done through text messaging.
Swatting	A cybercrime that involves terrorizing another person through making a false report using the Internet or smartphone that results in armed police showing up at an unsuspecting individuals residence.

CONCLUSION

It is difficult to imagine life without the Internet. It allows us to stay informed and connected with others. At the same time, being online increases the chance that an individual will become a victim of crimes ranging from ransomware to identity theft. The Internet has changed the parameters of traditional crimes; cybercrimes do not know geographical boundaries or time. The Internet also masks the identity of perpetrators making it difficult for law enforcement to gather evidence. Among the traditional crimes that have transitioned online are harassment, stalking, bullying and nonconsensual pornography, which form a typology of cybercrimes known as cyber aggression. The development of a typology of cybercrimes referred to as cyber aggression raises some important questions: "why do some states take a lead in combating these crimes, whereas others lag behind," "how can individuals protect themselves from being victimized through cyber aggression" and "how does government effectively respond to cyber aggression?"

This chapter laid the foundation for exploring these questions through presenting background information on crimes categorized as forms of cyber aggression. Chapter 2 will delve into the literature on victimology and discusses the relationship between sexual orientation and the likelihood of being victimized through bullying, harassment, and their cyber companions. Chapter 3 reviews the legislative response to cyber aggression, and Chapter 4 presents an overview of the literature on strategies being adopted at the school board level to limit the spread of cyberbullying. Chapter 5 provides an empirical analysis of the impact of state-level factors on policy adoption, and Chapter 6 ends the book with a discussion of possible action in fighting cyber aggression.

REFERENCES

Allanson, P., Lester, R., & Notar, C. (2015). A history of bullying. *International Journal of Education and Social Science*, *12*(2), 31–36.

Bauam, K., Catalano, S., Rose, K., & Rand, M. (2009). *Stalking victimization in the United States* (NCJ 224527). Bureau of Justice Statistics Special Report. Washington, DC: U.S. Department of Justice. Retrieved May 9, 2014, from http://www.bjs.gov/index.cfm?ty=pbdetail&iid=365

Burk, F. L. (1897). Teasing and bullying. *The Pedagogical Seminary*, *4*(3), 336–371. doi:10.1080/08919402.1897.10534145

Buxton, M. (2017, March 28). *Meet the congresswomen on the frontlines of the fight against online abuse*. Retrieved July 19, 2017, from http://www.refinery29.com/2017/03/147268/katherine-clark-online-abuse-interview

Capretto, L. (2016, April 6). *A father's painful crusade against bullying 12 years after his son's death*. Retrieved April 19, 2017, from http://www.huffingtonpost.com/entry/john-halligan-ryan-suicide_us_57043f13e4b0537661880e93

Catalano, S. (2012). *Stalking victims in the United States - revised.* Bureau of Justice Statistics Special Report (NCJ 224527). Retrieved July 12, 2014, from http://www.bjs.gov/content/pub/pdf/svus_rev.pdf

Citron, D., & Franks, M. (2014). Criminalizing revenge porn. *Wake Forest Law Review*, *49*, 345–391.

Cohn, D., & Caumont, A. (2016). *10 demographic trends that are shaping the U.S. and the world.* Retrieved March 18, 2017, from http://www.pewresearch.org/fact-tank/2016/03/31/10-demographic-trends-that-are-shaping-the-u-s-and-the-world/

Cyber Civil Rights Initiative. (2017). *What is 'revenge porn'?* Retrieved July 24, 2017, from https://www.cybercivilrights.org/revenge-porn-laws/

Cyberbullying Research Center. (2016). *Bullying laws across America.* Retrieved November 18, 2016, from http://cyberbullying.org/bullying-laws

D'Ovidio, R., & Doyle, J. (2003). A Study of Cyberstalking: Understanding investigative hurdles. *FBI Law Enforcement Bulletin*, *72*(3), 10–17.

Eckholm, E., & Zezima, K. (2010, March 29). *6 Teenagers are charged after classmate's suicide.* Retrieved December 9, 2014, from http://www.nytimes.com/2010/03/30/us/30bully.html?pagewanted=all&_r=0

Eisenberg, M. E., & Aalsma, M. C. (2005). Bullying and peer victimization: Position paper of the society for adolescent medicine. *The Journal of Adolescent Health*, *36*(1), 88–91. doi:10.1016/j.jadohealth.2004.09.004 PMID:15661606

Electronic Frontier Foundation. (2017). *Section 230 of the Communication Decency Act.* Retrieved December 27, 2017, from https://www.eff.org/issues/cda230

Ellis, R. (2017, December 31). *Swatting case poses legal challenges for police, prosecutors.* Retrieved January 3, 2018, from http://www.cnn.com/2017/12/31/us/swatting-legal-ramifications/index.html

Finkelhor, D., Turner, H. A., & Hamby, S. (2012). Let's prevent peer victimization, not just bullying. *Child Abuse and Neglect-the International Journal*, *36*(4), 271–274. doi:10.1016/j.chiabu.2011.12.001 PMID:22483362

Fisher, G. (2017, February 25). *High-risk, high expense security for celebrities plagued by stalkers*. Retrieved May 25, 2017, from http://www.cbsnews.com/news/high-risk-high-expense-security-for-celebrities-like-gwyneth-paltrow-sandra-bullock-plagued-by-stalkers/

France, L. (2017, July 10). *'Devastated' Blac Chyna granted restraining order against Rob Kardashian*. Retrieved July 15, 2017, from http://www.cnn.com/2017/07/10/entertainment/blac-chyna-gma-restraining-order/index.html

Frankel, T. C. (2012, October 20). *Megan Meier's mom is still fighting bullying*. Retrieved March, 12, 2014, from http://www.stltoday.com/news/local/metro/megan-meier-s-mom-is-still-fighting-bullying/article_f901d3e0-b6b8-5302-ac0c-80b83c9703a9.html

Gladden, R. M., Vivolo-Kantor, A. M., Hamburger, M. E., & Lumpkin, C. D. (2014). *Bullying surveillance among youth: Uniform definitions for public health and recommended data elements, version 1.0*. Atlanta, GA: National Center for Injury Prevention and Control, Centers for Disease Control and Prevention and U.S. Department of Education. Retrieved May 15, 2017, from http://www.cdc.gov/violenceprevention/pdf/bullying-definitions-final-a.pdf

Grenoble, R. (2012). *Amanda Todd: Bullied Canadian teen commits suicide after prolonged battle online and in school*. Retrieved January 10, 2017, from http://www.huffingtonpost.com/2012/10/11/amanda-todd-suicide-bullying_n_1959909.html

Hern, A. (2016, May 9). *How revenge porn sites rely on legal loopholes and anonymity*. Retrieved May 15, 2017, from https://www.theguardian.com/technology/2016/may/09/revenge-porn-websites-legal-loopholes-anonymity

Hertz, M., & David-Ferdon, C. (2008). *Electronic media and youth violence: A CDC issue brief for educators and caregivers*. Retrieved May 23, 2017, from https://www.cdc.gov/violenceprevention/pdf/ea-brief-a.pdf

Institute of Medicine &National Research Council. (2014). *Building capacity to reduce bullying: Workshop summary*. Washington, DC: The National Academies Press.

Interagency Working Group on Sexual Exploitation of Children. (2016). *Terminology guidelines for the protection of children from sexual exploitation and sexual abuse*. Retrieved July 22, 2017, from http://luxembourgguidelines.org/english-version/

Jany, L. (2016, June 8). *In electronic era, stalkers, abusers can be hard to track*. Retrieved May 12, 2017, from http://www.startribune.com/in-electronic-era-stalkers-abusers-can-be-hard-to-track/382278371/

Koo, H. (2007). A Time Line of the Evolution of School Bullying in Differing Social Contexts. *Asia Pacific Education Review*, *8*(1), 107–116. doi:10.1007/BF03025837

Leefeldt, E. (2016, May 10). *Victim of cyberbullying? Insurance might help*. Retrieved July 5, 2017, from http://www.cbsnews.com/news/victim-of-cyberbullying-insurance-might-help/

Limber, S. (2011). Development, evaluation, and future directions of the Olweus Bullying Prevention Program. *Journal of School Violence*, *10*(1), 71–87. doi:10.1080/15388220.2010.519375

Lonardo, T., Martland, T., & White, D. (2016). A legal examination of revenge pornography and cyber-harassment. *Journal of Digital Forensics, Security and Law*, *11*(3), 79–106. doi:10.15394/jdfsl.2016.1412

Mullen, P., & Pathe, M. (2002). Stalking. In M. Tonry (Ed.), *Crime and justice: A review of research* (Vol. 29, pp. 273–318). Chicago: University of Chicago Press.

Musu-Gillette, L., Zhang, A., Wang, K., Zhang, J., & Oudekerk, B. A. (2017). *Indicators of school crime and safety: 2016 (NCES 2017-064/NCJ 250650)*. Washington, DC: U.S. Department of Justice. doi:10.100710964-013-9925-5

National Academies of Sciences, Engineering, and Medicine. (2016). *Preventing Bullying Through Science, Policy, and Practice*. Washington, DC: The National Academies Press. doi: 10.17226/23482

National Conference of State Legislatures. (2015). *Cyberstalking and Cyber harassment Laws*. Retrieved September 12, 2015, from http://www.ncsl.org/research/telecommunications-and-information-technology/cyberstalking-and-cyberharassment-laws.aspx

National Network to End Domestic Violence, Safety Net Project, 2009. (2009). Retrieved September 15, 2015, from https://nnedv.org/latest_update/technology-facilitated-stalking/

National White Collar Crime Center. (2013). *Cyberbullying, the legislative response, NW3C research brief*. Retrieved May 11, 2017, from https://www.nw3c.org/docs/research/cyber-bullying.pdf?sfvrsn=6

NoBullying.com. (2015a). *Ryan Halligan loses his life to taunts, rumors and cyber bullying*. Retrieved July 11, 2015, from https://nobullying.com/ryan-halligan

NoBullying.com. (2015b). *Jessica Logan-the rest of the story*. Retrieved July 12, 2017, from https://nobullying.com/jessica-logan/

NoBullying.com. (2017). *The unforgettable Amanda Todd story*. Retrieved June 8, 2017, from https://nobullying.com/amanda-todd-story/

Ohlheiser, A. (2015, December 3). *Revenge porn purveyor Hunter Moore is sentenced to prison*. Retrieved July 18, 2017, from https://www.washingtonpost.com/news/the-intersect/wp/2015/12/03/revenge-porn-purveyor-hunter-moore-is-sentenced-to-prison/?utm_term=.4e6ac5f50832

Olweus, D. (1993). *Bullying at school: What we know and what we can do*. Oxford, UK: Blackwell.

Olweus, D. (2011). Bullying at school and later criminality: Findings from three Swedish community samples of males. *Criminal Behaviour and Mental Health*, *21*(2), 151–156. doi:10.1002/cbm.806 PMID:21370301

PACER National Bullying Prevention Center. (2017). *Questions answered*. Retrieved May 19, 2017, from http://www.pacer.org/bullying/

Pew Research Center. (2014). *Online harassment*. Retrieved February 16, 2015, from http://www.pewinternet.org/files/2014/10/PI_OnlineHarassment_102214_pdf1.pdf

Pew Research Center. (2017). *Online Harassment 2017*. Retrieved July 20, 2017, from http://www.pewinternet.org/2017/07/11/online-harassment-2017/

Powell, A. (2016). The sexting "epidemic": What can teachers do to prevent disruptions in school caused by sexting? *Journal of Law & Education*, *45*(4), 605–612.

Purcell, R., Pathé, M., & Mullen, P. E. (2012). Classification of stalkers. *Family & Intimate Partner Violence Quarterly*, *5*(2), 151–173.

Schnell, P., & Garcia, M. (2013). Connecting the dots: The challenges of identifying and responding to stalking. *The Police Chief*, *80*(12), 62–64.

Southworth, C., Finn, J., Dawson, S., Fraser, C., & Tucker, S. (2007). Intimate partner violence, technology, and stalking. *Violence Against Women*, *13*(8), 842–856. doi:10.1177/1077801207302045 PMID:17699114

Spence-Diehl, E. (2003). Stalking and technology: The double-edged sword. *Journal of Technology in Human Services*, *22*(1), 5–18. doi:10.1300/J017v22n01_02

Spitzberg, B., & Cupach, W. (2007). The state of the art of stalking: Taking stock of the emerging literature. *Aggression and Violent Behavior*, *12*(1), 64–86. doi:10.1016/j.avb.2006.05.001

Srabstein, J., & Merrick, J. (2013). *Bullying: A public health concern*. New York: Nova Publishers.

Stalking Resource Center. (2003). *Stalking technology outpaces state laws*. Retrieved March 2, 2014, from http://victimsofcrime.org/docs/src/stalking-technology-outpaces-state-laws17A308005D0C.pdf?sfvrsn=2

Stalking Resource Center. (2017). *What is stalking?* Retrieved June 19, 2017, from http://victimsofcrime.org/our-programs/stalking-resource-center/stalking-information

United Nations Children's Fund. (2014). *Hidden in plain sight: A statistical analysis of violence against children*. UNICEF. Retrieved May 25, 2017, from http://files.unicef.org/publications/files/Hidden_in_plain_sight_statistical_analysis_EN_3_Sept_2014.pdf

United Nations Educational, Scientific and Cultural Organization. (2017). *School Violence and Bullying*. Retrieved on December 12, 2017 from http://unesdoc.unesco.org/images/0024/002469/246970e.pdf

United States Department of Education. (2010). *Dear colleague letter*. Retrieved May 15, 2017, from http://www2.ed.gov/about/offices/list/ocr/letters/colleague-201010.html

United States Department of Health & Human Services. (2015). *What is cyberbullying?* Retrieved January 5, 2015, from http://www.stopbullying.gov/

United States Department of Health & Human Services. (2017a). *Prevent bullying*. Retrieved May 12, 2017, from https://www.stopbullying.gov/prevention/index.html

United States Department of Health and Human Services. (2017b). *Who is at risk? Considerations for specific groups*. Retrieved May 17, 2017, from http://www.stopbullying.gov/at-risk/groups/index.htmld

Vivolo, A., Holt, M., & Massetti, G. (2011). Individual and contextual factors for bullying and peer victimization: Implications for prevention. *Journal of School Violence*, *10*(2), 201–211. doi:10.1080/15388220.2010.539169

Wang, J., Iannotti, R. J., & Nansel, T. J. (2009). School bullying among adolescents in the United States: Physical, verbal, relational, and cyber. *The Journal of Adolescent Health*, *45*(4), 368–375. doi:10.1016/j.jadohealth.2009.03.021 PMID:19766941

Wolak, J., Mitchell, K., & Finkelhor, D. (2007a). Unwanted and wanted exposure to online pornography in a national sample of youth internet users. *Pediatrics*, *119*(2), 247–257. doi:10.1542/peds.2006-1891 PMID:17272613

Wolak, J., Mitchell, K., & Finkelhor, D. (2007b). Does online harassment constitute bullying? An exploration of online harassment by known peers and online-only contacts. *The Journal of Adolescent Health*, *41*(6), S51–S58. doi:10.1016/j.jadohealth.2007.08.019 PMID:18047945

Ybarra, M., Diener-West, M., & Leaf, P. (2007). Examining the overlap in internet harassment and school bullying: Implications for school intervention. *The Journal of Adolescent Health*, *41*(6), 42–50. doi:10.1016/j.jadohealth.2007.09.004 PMID:18047944

Chapter 2
Victimization:
Sexual Minorities

ABSTRACT

Not all groups are equally likely to be subject to acts of aggression; specific subgroups are more likely to be victimized. For example, youth who identify as a sexual minority are more likely to be victims of traditional forms of bullying than their heterosexual friends. There has been less research, however, on population subgroups and the likelihood of becoming a victim of cyber aggression. In exploring this topic, this chapter examines several questions including, "How important is the amount of time spent online as an intermediate variable in predicting whether an individual will become a victim of cyber aggression?" and "Does sexual orientation impact the likelihood of being a victim of cyberaggression above and beyond the amount of time spent online?" Multivariate statistical methods and survey data from the Pew Research Center for the year 2014 was used in this analysis.

INTRODUCTION

Within gender and sexual identity, students who identify as lesbian (L), gay (G), bisexual (B), transgender (T) or questioning (Q) (sexual minorities), are at a much higher risk for victimization than students who do not identify as a sexual minority. According to Kosciw et al., (2016), schools nationwide are hostile environments for a significant number of LGBTQ students, the overwhelming majority of whom routinely hear homophobic language and

DOI: 10.4018/978-1-5225-5285-7.ch002

experience victimization and discrimination while at school (p. XVI). As identified in the research literature, youths who are lesbian, gay, bisexual, or transgender report higher levels of physical, verbal, and relational aggression and bullying (PACER, 2017; U.S. Department of Health & Human Services, 2017a; Kosciw, et al., 2016; Kosciw, et. al., 2014; Gladden, et. al, 2014; Minton, 2014; GLSEN, CiPHR, & CCRC, 2013).

The media has played an important role in raising public awareness of cyber-aggression against certain vulnerable groups bringing attention to bullying of lesbian, gay, bisexual, transgender, and questioning persons. The media coverage of this issue began with Ryan Patrick Halligan who in 2003, at age 13, committed suicide after enduring online and in person taunts about his sexuality. Although there was no indication that Ryan was gay, the homophobic attack endured by Ryan raised awareness about the susceptibility to bullying faced by LGBTQ youth (NoBullying.com, 2015). While the death of Ryan Halligan brought public attention to bullying experienced by sexual minorities, it was the suicide of 18-year-old Rutgers student, Tyler Clementi in 2010, which helped to bring national and international attention to bullying and LGBTQ suicide. Tyler jumped off the George Washington Bridge in New Jersey after discovering that his roommate (Dharun Ravi) had set up a webcam in their dorm room that recorded Tyler engaging in an intimate relationship with another man. Ravi then posted the recording on Twitter (Spaulding, 2010; Leefeldt, 2016). Other examples include 14-year-old Kenneth Weishuhn who committed suicide in 2012 after being bullied at school and online and receiving death threats following his decision to "come out" as gay (Wong, 2012).

With each death like Ryan's, Tyler's and Kenneth's, there is often media coverage and heightened public awareness. The media, in reporting on these cases, is doing the job of agenda setting by sending a message to the public that bullying of sexual minorities is a topic of concern (Iyenger & Kinder, 1987). For public concern to develop into public policy, additional information, such as the scope of problem and the populations most likely to be impacted, is required. Thus, it becomes necessary to move beyond the initial media coverage and ask this question: to what extent are the LGBTQ populations more vulnerable than others? In exploring this topic, this chapter begins with an overview of the literature on sexual minorities and victimization.

A VULNERABLE POPULATION: SEXUAL MINORITIES

Sexuality occurs across a continuum which is to say that same-gender attraction and relationships are normal variations of human sexuality (American Psychological Association, 2015a; Substance Abuse and Mental Health Services Administration [SAMSHA], 2014; American Psychological Association, 2009; American Psychological Association, 2008). Likewise, a gender identity that is different than sex assigned at birth, and a gender expression diverging from the conventional cultural norms for gender, is normal variation of human gender (American Psychological Association, 2015a; SAMSHA, 2014). Sexual minorities include lesbian (L) women, gay (G) men, bisexual (B) men and women, and transgender (T) men and women. Questioning (Q) is a designation referring to individuals who are uncertain about their sexual orientation and/or gender identity. Q is also used as a verb to describe the process of exploring one's sexual orientation and/or gender identity (Buist & Lenning, 2016; SAMSHA, 2015).

The spectrum of sexuality includes orientation (attraction), expression (behavior), and identity (self-identification). Sexual orientation is based on the gender of the person or persons to whom someone is emotionally, physically, sexually, or romantically attracted and is expressed in relationship to others who fill a need for love, attachment, and intimacy (National LGBTI Health Alliance, 2015; SAMHSA, 2014). Sexual orientation is a multi-component construct commonly measured in three ways: attraction (e.g., the sex of a person one is sexually attracted to), behavior (e.g., ask respondents to report on the sex of people with whom they had willing sexual experiences), and self-identification (e.g., how would you describe your sexual orientation) (Gladden, et.al, 2014; Badgett & Goldberg, 2009; Saewyc et al., 2004). The word lesbian is generally used to describe women who are romantically and sexually attracted to other women (National LGBTI Health Alliance, 2015). The word gay is generally used to describe men who are romantically and sexually attracted to other men. Gay is sometimes used to refer to the general LGBT community. Bisexuality is the capacity for emotional, romantic, and/or physical attraction to more than one gender. A bisexual orientation speaks to the potential for, but not requirement of, involvement with more than one gender (GLSEN, 2017). Cisgender is a term denoting an individual whose gender identity or expression is consistent with his/her sex assigned at birth (GLSEN, 2017). The term gender minority refers to transgender and gender nonconforming people whose gender identities or gender expressions fall outside of the social norms typically associated with their assigned sex at birth (Reisner, et.al, 2015; Hendricks & Testa, 2012). The transgendered population is a highly vulnerable group among sexual minority populations; gender minority youth experience high rates of bullying, harassment, and other types of peer victimization (Lambda, 2017; Buist & Lenning, 2016; Kosciw et. al., 2012).

VICTIMIZATION: SEXUAL MINORITIES

National Center for Education Statistics and the Bureau of Justice Statistics (Musu-Gillette, Zhang, Wang, Zang, & Oudekerk, 2017)

Indicators of School Crime and Safety: 2016 is the 19th in a series of reports produced since 1998 by the National Center for Education Statistics (NCES) and the Bureau of Justice Statistics (BJS) that present the most recent data available on school crime and student safety. The indicators in the report are based on information drawn from a variety of independent data sources, including national and international surveys of students, teachers, principals, and postsecondary institutions and universe data collections from federal departments and agencies and international organizations. The sources include Bureau of Justice Statistics, National Center for Educational Statistics, the Federal Bureau of Investigation, the Centers for Disease Control and Prevention, the Office of Postsecondary Education, the Office for Civil Rights, and the International Association for the Evaluation of Educational Achievement. Each data source has an independent sample design, data collection method, and questionnaire design, or is the result of a universe data collection. (Musu-Gillette, et al., 2017).

According to the report, sexual minority youth are at a greater risk of harassment, victimization, and social isolation as compared to heterosexual youth. Experiences such as these can result in symptoms of depression, suicidal thoughts, problem behaviors, inferior academic outcomes and increased unexcused absences from school. In 2015, a higher percent of gay, lesbian, or bisexual students (34%) in grades 9–12 than of heterosexual students (19%) reported having been bullied on school property during the previous 12 months. The percent of students reporting being bullied on school property was also higher for students who were not sure about their sexual orientation than for heterosexual students (25% vs. 19%). Likewise, a higher percent of gay, lesbian, or bisexual students reported being electronically bullied during the previous 12 months in 2015 than did heterosexual students, (28% vs. 14%). The percent of students who reported being electronically bullied was also higher for students who were not sure about their sexual orientation than for heterosexual students, overall (23% vs. 14%) (Musu-Gillette, et al., 2017).

GLSEN (Kosciw, Greytak, Giga, Villenas, & Danischewski, 2016)

The 2015 National School Climate Survey: The experiences of lesbian, gay, bisexual, transgender, and queer youth in our nation's schools (2016) was conducted online. Efforts to ensure a representative national sample of lesbian, gay, bisexual, transgender, and queer (LGBTQ) youth, included outreach through national, regional, and local organizations providing services to or advocated on behalf of LGBTQ youth, and advertisements on social networking sites, such as Facebook, Instagram, and Tumblr. To ensure representation of transgender youth, youth of color, and youth in rural communities, efforts were made to notify groups and organizations that worked predominantly with those populations. The final sample consisted of a total of 10,528 students between the ages of 13 and 21years. Students were from all 50 states and the District of Columbia and from 3,095 unique school districts. About two-thirds of the sample (68.6%) was White, a third (34.9%) was cisgender female, and about half identified as gay or lesbian (49.2%). Students were in grades 6 to 12, with the largest numbers in grades 10 and 11.

According to the results of the survey, almost 60% of LGBTQ students reported feeling unsafe at school because of their sexual orientation; 40% reported feeling unsafe at school because of how they expressed their gender. Almost one third of students missed at least one day of school in the past month because they felt unsafe or uncomfortable. LGBTQ students reported regularly avoiding school bathrooms and locker rooms because they felt unsafe or uncomfortable in those spaces. Most LGBTQ students reported avoiding school functions and extracurricular activities to some extent, and about 25% avoided them often or frequently (Kosciw et al., 2016).

Sexual orientation and gender expression were the most common reasons LGBTQ students were harassed or assaulted at school with approximately 90% of respondents indicating they were harassed at school. Nearly 75% of students reported being verbally harassed at school because of their sexual orientation; more than 50% were verbally harassed because of their gender expression. Over 25% of students reported being physically harassed at school because of their sexual orientation; 1 in 5 were physically harassed because of their gender expression. About 1 in 6 students reported being physically assaulted at school in the past year, primarily because of their sexual orientation, gender expression, or gender. Overall, transgender students were more likely than all other students to have negative experiences at school. Transgender students were more likely to have felt unsafe and to experience victimization at school based on their gender expression or gender (Kosciw et al., 2016).

Center for Disease Control (Kann, Olsen, McManus, et al. 2016)

The Center for Disease Control, in August 2016, released the first nationally representative study on the health risks of U.S. lesbian, gay, and bisexual (LGB) high school students. The study examined bullying that occurred on school property and bullying that occurred online. The findings for bullying on school property indicated that 20.2% of all students; 18.8% of heterosexual students; 34.2% of gay, lesbian, and bisexual students; and 24.9% of "not sure" students had been bullied on school property during the 12 months before the survey. The prevalence of having been bullied on school property was higher among gay, lesbian, and bisexual students (34.2%) than not sure students (24.9%) and heterosexual students (18.8%). Among female students, the incidence of bullying was higher among lesbian and bisexual students (37.2%) than heterosexual students (23.2%) and not sure students (19.1%). Among male students, the prevalence was higher among gay and bisexual students (26.3%) and not sure students (31.7%) than heterosexual students (15.0%). As with bullying on school property, the chance of cyberbullying was higher among gay, lesbian, and bisexual students (28.0%) and not sure students (22.5%) than heterosexual students (14.2%). Among female students, the occurrence was higher among lesbian and bisexual students (30.5%) than heterosexual students (20.6%) and not sure students (20.4%). Among male students, the prevalence was higher among gay and bisexual students (22.4%) and not sure students (22.3%) than heterosexual students (8.7%). (Kann, et al.,2016)

Gay, Lesbian, and Straight Education Network (GLSEN) and Center for Innovative Public Health Research (CiPHR) & Crimes Against Children Research Center, University of New Hampshire (2013)

Out Online: The Experiences of Lesbian, Gay, Bisexual and Transgender Youth on the Internet (2013) was published by GLSEN (Gay, Lesbian & Straight Education Network) and provides an in-depth account of how LGBT young people navigate a space that can be both a critical lifeline and a site of vulnerability. Data used in this study come from the *Teen Health & Technology* survey conducted by Harris Interactive Inc. on behalf of the Center for Innovative Public Health Research (CiPHR), GLSEN, and the Crimes against Children Research Center at the University of Hampshire. The survey was conducted online between August 2010 and January 2011, with a total sample of 5,680 youth ages 13-18. According to the survey, LGBT youth were nearly three times as likely as non-LGBT youth to say they had been bullied or harassed online (42% vs. 15%) and twice as likely to say they had been bullied via text message (27% vs. 13%). Additionally, LGBT youth were as likely to report feeling unsafe online (27%) as they were at school (30%) and while traveling to and from school (29%). White LGBT youth experienced greater levels of online and text-based bullying and harassment than LGBT youth of other races/ethnicities. LGBT youth in rural areas experienced substantially higher levels of victimization online and via text message compared to LGBT youth in suburban and urban areas.

Summary of Research: Sexual Minorities

Children who are bullied are often marginalized by their peers for a wide variety of reasons; no single factor necessarily puts a child at risk (PACER, 2017; Department of Health & Human Services, 2017b). However, some groups, LQBT populations for one, are at an increased risk of being bullied (PACER, 2017; Department of Health & Human Services, 2017b). Research has documented that, compared to same-age heterosexual peers, LGBT youth are at increased risk of interpersonal violence, including forms of bullying, harassment, stalking, and their cyber companions and particularly within the school environment (Greytak, et. al., 206; Kosciw, et. al., 2016; Kann, et al., 2016; United Nations Children's Fund, 2014; GLSEN, CiPHR, & Crimes Against Children Research Center, 2013.)

While research on transgender youth is limited, we do know this population is highly victimized. In 2015, the National Center for Transgender Equality administered an anonymous online survey examining the experiences of transgender people (18 and older) in the United States (James, et. al., 2016). Participating in the survey were 27,715 respondents from all fifty states, the District of Columbia, American Samoa, Guam, Puerto Rico, and U.S. military bases overseas. The U.S. Transgender Survey (USTS) was an anonymous, online survey for transgender adults (18 years and older) in the U.S. In reflecting upon their experiences in schools, 77% of those who were out or perceived as transgender at some point between Kindergarten and Grade 12 (K-12) experienced some form of mistreatment, such as being verbally harassed, prohibited from dressing according to their gender identity, disciplined more harshly, or physically or sexually assaulted because people thought they were transgender. Fifty-four percent (54%) of those transgender in K–12 were verbally harassed, 24% were physically attacked, and 13% were sexually assaulted. Seventeen percent (17%) faced such severe mistreatment as a transgender person that they left a K–12 school. Additionally, 24% in college or vocational school were verbally, physically, or sexually harassed (Lambda Legal, 2017; Kosciw et. al., 2016; James, et.al, 2016; Kosciw et. al.,2012).

BULLYING AND CYBERBULLYING: THE IMPACT ON THE VICTIMS?

Bullying is linked to many negative outcomes such as substance abuse, mental health issues, and suicide. According to the Department of Health and Human Services, children and adolescents who are bullied are more likely to experience depression, anxiety, increased feelings of sadness and isolation, changes in sleep and eating patterns, and loss of interest in activities (2015). Additionally, bullied youth are more likely to be truant from school and/or to drop out of school (2015). A study on the relationship between bullying and suicide conducted by the Centers for Disease Control and Prevention(CDC) found that negative outcomes of bullying (youth who bully others, youth who are bullied, and youth who bully others and are bullied) may include depression, anxiety, involvement in interpersonal violence or sexual violence, substance abuse, poor social functioning, and poor school performance for example lower grade point averages, sub-standard test scores, and poor attendance (CDC, 2014a). Furthermore, youth who report frequent bullying of others and youth who report being frequently bullied are at increased risk for suicide-related behavior (CDC, 2014a).

BROAD SPECTRUM RESEARCH: STALKING AND CYBERSTALKING: WHAT DO WE KNOW IN GENERAL?

Discussions about research on the prevalence of cyberstalking began in earnest in the late 1990's when U.S. Attorney General Janet Reno identified cyberstalking as an emerging social problem (Fisher, Reyns, & Sloan, 2016; Reno, 1999). *Cyberstalking: A New Challenge for Law Enforcement and Industry: A Report From the Attorney General to the Vice President* outlined a ten step approach for responding to the crime of cyberstalking with respect to its nature and extent; the current approach that law enforcement, industry, victim groups, and others were using to address the problem; the adequacy of Federal and State laws; and recommendations to improve and enhance efforts to address the problem (Reno, 2000; Reno, 1999).

Currently, within the literature there is consensus that research on the crimes of stalking, cyberstalking and cyber-harassment is lacking (Catalano, 2012). Moreover, crimes of this type tend to be underreported which further obscures our ability to understand the prevalence of this type of behavior (National White Collar Crime Center, 2013). A common reason for not reporting the crime is the victim classifies the action(s) as a private and/or family matter. Additionally, the victim may not recognize the behavior as criminal (Schnell & Garcia, 2013). Actions such as gift giving, sending letters or email, rumormongering, and appearing at a site where the victim is present, are not in and of themselves necessarily criminal. Legislatively, it is the repetitiveness of the acts; a course of conduct that causes the victim to be fearful for personal or family safety that classifies many of the behaviors associated with stalking as illegal (Schnell & Garcia, 2013; Baum, Catalano, Rand, & Rose, 2009). Although these crimes often go unreported, impacting our research capabilities, there have been several major studies that provide an understanding of pervasiveness and underlying causes. Presented below is a summary of their findings.

National Crime Victimization Survey: Supplemental Study on Stalking (SVS) (2006)

In 2006, The Department of Justice's Office on Violence Against Women funded a supplemental study (SVS) to the National Crime Victimization Survey (NCVS) focusing on the crimes of non-fatal stalking, harassment and their cyber companions. The population consisted of all survey respondents over 18 years of age who experienced one of seven types of stalking behaviors on at least two occasions (Bureau of Justice Statistics, 2014). Additionally, the respondents must have experienced fear for personal and/or family safety (Catalano, 2012, p. 1). The seven behaviors that were measured in the study included receiving unwanted phone calls; receiving unwanted or unsolicited emails or letters; a perpetrator following or spying on the victim; a perpetrator appearing in places without a legitimate purpose; a perpetrator waiting for the victim at different locations; a perpetrator leaving unwanted items, gifts, or flowers; and the perpetrator posting information or spreading rumors about the victim on the internet, in a public place, or by word of mouth (Catalano, 2012, p.1). In addition, the survey included a measurement for harassment. Harassment was said to be present in cases where the respondent was victimized by one of the seven stalking behaviors, but for which fear or concern for personal safety was not present (Catalano, 2012, p.1).

Findings from the 2006 SVS indicated that approximately 14 out of every 1,000 persons age 18 or older, were victims of stalking (Baum, Catalano, Rand, & Rose, 2009, p.1). This equates to approximately 3.4 million people being stalked in the United States during a 12-month period extending over calendar years 2005 and 2006 (Baum, Catalano, Rand, & Rose, 2009, p.1). Women (2.2%) were more likely to be stalked than men (0.8%); however, both genders were equally likely to be victims of harassment. About 1 out of 4 or 25% of stalking victimization involved some form of cyberstalking such as unwanted emails or text messages. Nearly 70% of stalking victims were acquainted, at varying levels, with the perpetrator of their victimization. About 28% of the perpetrators were known current intimates such as spouse and significant others; 20% were former intimates such as ex-spouse and ex-significant others; about 42% were known others such as friends, roommates, relatives, and acquaintances; 9% were stranger; and 14% were unknown. Persons of younger age were more likely to be stalked than were persons of older age. Individuals most likely to be stalked included those in the age groups of 18 and 19 (2.9%) and 20-24 years (2.8%) (Catalano, 2012, p.4).

People in lower income brackets were at greater risk of being stalked than were those in higher income brackets. Over 6% of respondents with a household income of $14,999 or less reported being victimized; decreasing to 4% for a household income of $34,000 to $14,999; about 3% for a household income of $74,999 to $35,000; and 1% for a household income of $75,000 or more. Relative to marital or otherwise status, the results indicated that the risk of stalking victimization was the highest for individuals who were divorced or separated (3.3%) as compared to those who were never married (2.4%), widowed (1.4%) or married (0.8%) (Catalano, 2012, p.4).

The most prevalent stalking behavior reported was receiving unwanted phone calls and messages. Approximately 67% of all respondents received unwanted phone contacts and approximately 31% reported receiving unwanted letters and email messages. About 36% of the respondents were victimized through rumors and about 34% reported that the perpetrator followed or spied on them. A little less than ⅓ of the respondents (32%) reported that the perpetrator showed up at events without reason and about 30% reported that the perpetrator was waiting for them without reason. Finally, about 13% of the respondents reported receiving unwanted gifts, flowers, or other items (Catalano, 2012, p.4).

The National Intimate Partner and Sexual Violence Survey (NISVS) (Black, Basile, Breiding, et al., 2011)

The National Intimate Partner and Sexual Violence Survey is an on-going, nationally representative telephone survey that collects information about experiences of sexual violence, stalking, and intimate partner violence among non-institutionalized English and/or Spanish speaking women and men aged 18 and older in the United States. The study is conducted by the Center for Disease Control's National Center for Injury Prevention and Control with the support of the National Institute of Justice and the Department of Defense. The 2010 Summary Report was the first year of NISVS data on the national prevalence of intimate partner violence (IPV), sexual violence (SV), and stalking trends (Black, et al., 2011).

NISVS (2011, p. 29) found that approximately 1 in 6 women (16%) and 1 in 19 men (5%) have experienced stalking in their life time. Corresponding with the results from the SVS, this survey found that the majority (66%) of female victims were stalked by current or former intimate partners. Male victims were stalked by current or former intimate partners (41%) or acquaintances (40%). Receiving unwanted phone, email, and text messages were the most common tools of stalking for female victims (79%) as well as for male victims (76%). Furthermore, more than 50% of female victims and more than 33% of male victims indicated that they were stalked before the age of 25 (NISVS, 2013, p 43). Finally, summary results from the survey indicated that IPV, SV, and stalking were widespread in the U.S. Women were found to be disproportionately affected by IPV, SV, with stalking resulting in health concerns that were identified by CDC as lifelong and inclusive of medical (long term chronic disease) and psychological (post-traumatic stress disorder) consequences (Breiding, Chen & Black, 2014, p. 61).

Other Studies: The First National Study on Stalking and A Population of College Students

In the first-ever national study on stalking, Tjaden & Thoennes (1998, p.2) found that 8% of women and 2% of men had been stalked in their lifetime; an annual estimate of 1,006,970 women and 370,990 men. While stalking is a gender-neutral crime, most victims identified in this survey were female (78%); four out of five stalking victims were women. American Indian/Alaska Native women were more likely to report being victimized than women of other racial or ethnic backgrounds. Young adults between the ages of 18 and 29 years were the mostly likely to be stalked (52%). Seventy seven percent (77%) of the women in the sample knew their stalker and of that number 59% were stalked by a former significant other. Sixty-four percent (64%) of males knew their stalker and of that number 30% were stalked by a former significant other (1998, pp. 5-6). The data used in this research were from the National Violence Against Women Survey (NVAW) sponsored by the National Institute of Justice and Centers for Disease Control and Prevention and conducted by the Center for Policy Research. Data were collect from November 1995 through May 1996.

Reyns, Henson, & Fisher (2012, pp. 11-12) found about 41% of a random sample of college undergraduate students at an urban Midwest university had experienced cyberstalking in their lifetime, which is a larger percentage than found in other studies. Consistent with survey data from SVS and NISVS, females (46%) were more likely to be victimized than males (32%). About 17% of those victimized reported their experience to law enforcement officials. Unlike other studies and surveys, Reyns, Henson, and Fisher (2012, p. 15) found that the perpetrator in most incidents was unknown or a stranger to the victim (44%); or a friend or acquaintance (39%); or an intimate (17%). This finding could be indicative of the sample from which the data was drawn (large urban college campus in the Midwest) in comparison to national survey data that draws from a far more diverse population sample. In general, this research identified females, nonwhites, non-heterosexuals, and non-singles as experiencing higher rates of cyberstalking victimization than males, whites, heterosexuals, and singles.

A Summary of the Research: Logan (2010)

Logan (2010) conducted a general overview of peer reviewed and published research on stalking. This meta-analysis included research reports from the National Institute of Justice and similar organizations such as the Stalking Resource Center and the National Network to End Domestic Violence. This project was supported by the National Institute of Justice. In general, the research indicates that intimate partner stalking is a relatively common type of violence against women and comprises the largest category of stalking cases. Between 4.8% and 14.5% of women (ages 18 and older) report being stalked and victimized by an intimate partner during their lifetime. Numerous studies identified a relationship between partner stalking and a history of partner physical violence, sexual violence, and coercive control. The tactics or tools used to stalk varied considerably (Logan, 2010). Physical surveillance such as being followed, spied on, or watched and unwanted phone and other types of contacts such as email, gifts, text messages, and letters were frequently cited in the research. Few studies specifically addressed the use of technology to stalk or cyberstalk. One study found relatively low rates of technology use (25%) in comparison with stalking in general (Baum, Catalano, Rand, & Rose, 2009). A second study found the use of email and/or Internet in 15% of the cases (partner stalking) and all other types of technology excluding GPS in 13% of the cases (Botuck, Berretty, Cho, Tax, Archer, & Cattaneo, 2009). In general, stalking victims exhibit higher levels of fear and distress and stalking victimization is associated with a range of fears and significant psychological distress (Logan, 2010). Partner stalking was also found to be associated with elevated sleep and health problems.

Finally, there was limited information in the research on the rate that stalking cases are reported to the justice system. Estimates from a few studies indicated between 52% and 72% of the victims had some type of contact with law enforcement (Logan, 2010, p. 15). Victims cited several reasons for not reporting stalking behavior to the justice system including fear that no one will believe them; a belief that the consequences of reporting the behavior are too severe; a belief that the police could not or would not do anything; fear of the stalker; lack of proof; and a belief that the event was a private matter (Logan, 2010). Aggregated, most studies identify the crimes of stalking and cyberstalking as gender specific with women being victimized in significantly greater numbers than male victims. Research, except for the Reyns, Henson & Fisher study (2012), found that victimized females were most likely acquainted on some level with the perpetrator of their victimization. Younger, divorced or separated women, in lower income brackets were at greatest risk of victimization. Survey data from SVS and NISVS identified unwanted messages and phone contacts as the most prevalent stalking behavior identified by respondents. Numerous studies in the meta-analysis identified a relationship between partner stalking and a history of partner physical abuse, sexual violence, and coercive control (Logan, 2010, p. 4). Finally, data from the SVS found the prevalence of cyberstalking was about 25%, or one out of four respondents.

TIME SPENT ON-LINE AND SEXUAL MINORITIES

The current literature suggests that sexual minorities as more likely to be victims of bullying, stalking and harassment. There has been less research, however, on this population and cyber aggression. This raises several questions. How important is the amount of time spent online as an intermediate variable? Does sexual orientation impact the likelihood of being a victim of cyber aggression above and beyond the amount of time spent online? In the next section, these questions will be examined using multivariate statistical methods. More specifically, it will examine the impact of sexual orientation and the time spent online on being a victim of cyberaggression.

EMPIRICAL MODEL: DATA AND METHODS

Data

In examining these questions, this study uses the *2014 Teen Relationship Survey* conducted for the Pew Research Center by GfK Group (formally known as the Knowledge Network). The Pew Research Center is a nonpartisan, nonprofit group that conducts studies and provides information regarding factors that shape American society. Participants of this survey were asked questions regarding online relationships. This study is based on a sample size (n) of 1,060 teens (ages 13 to 17) living in the United States, collected from September 25, 2014 to October 9, 2014 and from February 10, 2015 to March 16, 2015. Data was collected using a probability-based survey conducted online (Lenhart, Smith, & Anderson, 2015).

Cyberbullying and Sexual Orientation

Bullying/cyberbullying can take on many forms ranging from spreading rumors to attacking someone physically or verbally (U.S. Department of Health & Human Services, 2015). This study explores two forms of online bullying; purposely excluding someone from a group and the spreading of rumors. It explores the relationship between cyberbullying and sexual orientation while controlling for other individual factors identified in the literature related to the likelihood of being bullied. Two dependent variables were used. The first is as a dummy variable constructed using the following question: "Do you ever experience any of the following on social media: people posting about things you weren't invited to?" This question is used as a measure of social exclusion. The second dependent variable is a dummy variable constructed using the following question: "Do you ever experience any of the following on social media: people posting things about you that you can't change or control?" The variable acts as a measure of rumor spreading. Both variables are coded 1 if an individual responded "yes" and 0 otherwise.

There are two main independent variables; the first is sexual orientation, which is measured using a binary variable coded 1 if the respondent self identifies as a for sexual minority and 0 if the identify as a heterosexual. While the question is a measure of sexual orientation, it does not indicate whether the respondent has "come out" to their friends and classmates. This limit, to some extent, how the findings can be interpreted. The second main independent variable is an indication of how often a respondent uses the Internet, which is measured on a 6-point scale where 1 indicates less than once a week and 6 indicates almost constantly. Control variables include an income measure on a 9-point scale where 1 indicates that family incomes range from $0 to $10,000 and 9 signifies a family income of $150,000 or more. Zweig et al. (2013) found that among the individuals more likely to experience cyberbullying are younger, females and Whites. To control for the influence of gender a binary variable coded 1 for male and 0 for female is included. To control for race and ethnicity, dummy variables were included for African Americans, Latinos and other with non-Hispanic Whites as the reference group. Age (between 13 and 17) is measured in years. Dummy variables for the South, Northeast and West regions were included to control for regional differences; survey findings indicate that the rate of cyberbullying are higher in the Midwest and Northeast (U.S. Department of Education, 2013, p. T-23).

FINDINGS AND DISCUSSION

In Table 1, the dependent variables were coded so that higher scores are associated with increased likelihood of being cyberbullied. Since the dependent variables are binary, the models are estimated using logistic regression.

Table 1. Sexual orientation and cyberbullying, 2014

Variables	Social Exclusion		Rumor Spreading	
	β (se)	p>|z|	β (se)	p>|z|
Environmental Variables				
South	.150(.199)	.452	-.090(.200)	.653
Northeast	.377(.244)	.122	.216(.238)	.365
West	.102(.226)	.653	.197(.226)	.383
Individual Level Variables				
Age	.053(.053)	.321	**.110(.053)**	**.038**
Male	**-.323(.149)**	**.031**	-.123(.150)	.411
Latino	**-.469(.206)**	**.023**	**-.719(.213)**	**.001**
Black	-.413(.263)	.117	-.532(.270)	.109
Other	-.084(.261)	.748	-.167(.257)	.516
Family Income	.022(.017)	.193	.015(.018)	.390
Sexual Orientation	**.641(.109)**	**.022**	**.502(.268)**	**.061**
Internet Usage	**.425(.109)**	**.000**	**.197(.108)**	**.066**
Constant	**-2.927(.983)**	**.003**	**-3.000(.986)**	**.002**
Pseudo R2	.0383		.0297	
LR Chi² (11)	40.41	.038	31.06	.001
N	767		767	

Sources: *2014 Teen Relationship Survey.* Logistic regression estimates with standard errors in parentheses. Reported probabilities are based on two-tailed tests. Statistically significant coefficients at .10 or less in bold.

As predicted by the literature, both forms of cyberbullying examined in this study (social exclusion and rumor spreading) were found to be positively associated with the increased Internet use and being a person who identifies as LGBTQ. Furthermore, the dependent variables are based on questions regarding social media use. Prior research (Mitchell, Finkelhor, & Wolak, 2003; Wolak, Mitchell, & Finkelhor, 2007; Marcum, 2008) argue that it is not simply the amount of time spent online that increases the chance of being bullied, but also the specific online activities such as communicating with others and sharing personal information. Since the dependent variables concern bullying based on social media use, they also support this earlier research. Other results were not consistent with previous research. Neither regional indicators nor family income were found to be associated with either form of bullying and there was limited support for the argument that race/ethnicity, gender and age impact the chance of being cyberbullied. Partially supporting previous findings is that Whites are more likely to be cyberbullied. In both models, Latinos were found to be less likely to be cyberbullied then non-Hispanic Whites. Otherwise, race/ethnicity was found unrelated to being bullied online. In addition, females were more likely to be victims of social exclusion while older teenagers were more likely to be victims of gossiping. These findings suggest that future surveys should distinguish between different forms of cyberbullying. Consistent with research on traditional bullying, certain populations (e.g. females) are more likely to be victims but it is conditional on the type of cyberbullying.

MONTE CARLO SIMULATION

To further explore the relationship between sexual orientation and bullying online, the percentage of individuals who were bullied online from Table 1 is presented in Table 2 and 3. While the findings in Table 1 show both sexual orientation and the extent of Internet usage are predictors of the chance of being bullied online, it is unlikely that the impact is the same for every form of cyberbullying. To illustrate the extent to which online usage and sexual orientation can predict being cyberbullied, the coefficients reported in Table 1 were converted to predicted probabilities of being bullied online (King, Tomz, & Wittenberg, 2000). The simulations compare the probability of being bullied online between sexual minorities and heterosexuals. Probability simulations were calculated holding the variables age and family income at their medians. Values for gender was set at female, race at non-Hispanic white and region at Midwest. Expected probabilities were calculated for different levels of Internet usage.

Sexual Orientation and Likelihood of Being Cyberbullied (Social Exclusion)

Table 2 presents the predicted probabilities for being a victim of cyberbullying (socially excluded) based on sexual orientation and for different levels of Internet usage. The probability of being excluded increases between 11.7% and 14% for sexual minorities when compared to heterosexuals. Although these findings indicate that identifying as LGBTQ increases your risk of being bullied online by approximately 12%, the amount of time spent online significantly increases this risk. For sexual minorities the likelihood of being bullied increases from 32.2% to 78.8% (a difference of 46.6%) when the amount of time spent online increases from less than once a week to constantly being online. Similarly, for heterosexuals the likelihood of being bullied increases from 20.5% to 67% (a difference of 46.5%). These findings support previous research that finds individuals who identify as part of the LGBTQ community are more likely to be victims of cyberbullying, but in making predictions, the amount of time spent online, and online activities need to be factored into the equation.

Table 2. The relationship between sexual orientation and likelihood of being cyberbullied (social exclusion)

Internet Use/Sexual Orientation	Low	Median	High
Sexual Minority	32.2% (10.90)	71.0% (6.24)	78.8% (5.49)
Heterosexual	20.5% (7.80)	57.0% (4.21)	67.0% (4.39)
Difference (Sexual Minority-Heterosexual)	11.7%	14.0%	11.8%

Note: Standard errors are in parentheses. To simulate different levels of Internet use, usage was set at less than once a week, median, and almost constantly. Sexual orientation is set at sexual minority and heterosexual. Value for race was set at White, family house income at median, age at median, gender at female and region at the Midwest. Estimations were produced using Clarify: Software for Interpreting and Presenting Statistical Results, by Michael Tomz, Jason Wittenberg, and Gary King (2000).

Sexual Orientation and Likelihood of Being Cyberbullied (Rumor Spreading)

Table 3 presents the predicted probabilities for being a victim of cyberbullying (rumor spreading or being gossiped about) based on sexual orientation and for different levels of Internet usage. The probability of being excluded increases between 10.8% and 12.2% for sexual minorities when compared to heterosexuals. These probabilities, while similar, are slightly less than those found for social exclusion. The findings show that identifying as LGBTQ increases your risk of being bullied online by approximately 11% and like the findings in Table 2 the amount of time spent online significantly increases this risk. For sexual minorities the likelihood of being bullied increases from 39.6% to 62.8% (a difference of 23.2%) when the amount of time spent online increases from less than once a week to constantly being online. Similarly, for heterosexuals the chance of being bullied increases from 28.8% to 51.2% (a difference of 22.4%). These findings are significantly less than those found in Table 2 and suggest that the Internet may facilitate certain forms of bullying more easily than others.

FUTURE RESEARCH DIRECTIONS

Secondary data analysis was used in this chapter to explore the relationship between online bullying and sexual orientation. Like all forms of data collection, this method has advantages and disadvantages. One disadvantage associated with this method is that the data was collected by someone else for their needs. While this dataset allowed us to examine the relationship

Table 3. The relationship between sexual orientation and likelihood of being cyberbullied (rumor spreading)

Internet Use/Sexual Orientation	Low	Median	High
Sexual Minority	39.6% (11.7)	58.2% (7.24)	62.8% (7.33)
Heterosexual	28.8% (9.26)	46.0% (4.28)	51.2% (4.75)
Difference (Sexual Minority-Heterosexual)	10.8%	12.2%	11.6%

Note: Standard errors are in parentheses. To simulate different levels of Internet use, usage was set at less than once a week, median and almost constantly. Sexual orientation is set at sexual minority and heterosexual. Value for race was set at White, family house income at median, age at median, gender at female and region at the Midwest. Estimations were produced using Clarify: Software for Interpreting and Presenting Statistical Results, by Michael Tomz, Jason Wittenberg, and Gary King (2000).

between sexual orientation and two forms of cyberbullying, future studies should explore other types of online bullying. In addition, this study was limited by age (13-17 years) and while the respondents were asked about their sexual orientation, they were not asked if they had "come out" to other such as friends, family and classmates. This additional information, could have added an addition dimension to the analysis. The dataset is also a relatively small size (n=1060). As a final issue, this study relies on data from a single time frame (a cross-sectional design), suggesting outcomes only relevant for a specific period. As the Internet continues to change, so does the chance that cyber aggression will evolve and the chance that an individual or specific population will become a victim. The results of a cohort study might identify factors that are consistently associated with online bullying or other forms of cyber aggression, while a panel study could illustrate the chance of being victimized at different stages of a person's life.

An additional avenue for expanding the research in this chapter, would be to include other subgroups that are more likely to be to victims of cyber aggression. For example, research on traditional forms of aggression find that along with sexual minorities, persons with disabilities and special health needs are more likely to be victims of bullying, stalking and harassment. The literature (Langevin, Bortnick, Hammer, & Wiebe, 1998; Finkelhor, 2008; Rose, Monda-Amaya, & Espelage, 2010) finds that bullying of youth and teens is higher among those with a disability. Women and girls with disabilities are also at a higher risk of sexual assault (Sobey, 1994; Sullivan & Knutson, 2000; Rand & Harrell, 2009). Research on the relationship between cyber aggression and people with disabilities however is limited. Kowalski and Fedina (2011) found that students with ADHA or Asperger syndrome experienced cyberbullying at a rate lower than traditional bullying. In addition, Jovonen, and Gross (2008) found that students with disabilities experienced less cyberbullying than traditional bullying because they were less likely to go online. This work is older and limited and more up-to-date research is needed to determine if these results still hold.

CONCLUSION

This chapter represents a preliminary effort to examine the extent that sexual minorities are more likely to be targets of cyber aggression. The literature on victimization points to the fact that persons who identify as LGBTQ (sexual minorities) are more likely to be targeted than persons who do not identify as a sexual minority. The literature on cyberbullying (Mitchell, Finkelhor, & Wolak, 2003; Wolak, Mitchell, & Finkelhor, 2007; Marcum, 2008; Jones et al., 2013) finds the greater the amount of time spent online along with certain online activities, the greater the chance of being victimized. The statistical models in this chapter examined the relationship between sexual orientation and two forms of online bullying (social exclusion and rumor spreading). The models found that being a sexual minority and the amount of time spent online were positively associated with both forms of online bullying. Monte Carlo simulations suggest sexual minorities were approximately 12% more likely to be socially excluded and 11% more likely to be gossiped about than heterosexuals, online. The findings also show that online activities and the amount of time an individual spends online does influence the chance that a person will be cyberbullied.

REFERENCES

American Psychological Association. (2008). *Answers to your questions: For a better understanding of sexual orientation and homosexuality*. Washington, DC: Author. Retrieved May 22, 2017, from www.apa.org/topics/lgbt/orientation.pdf

American Psychological Association. (2015). *Guidelines for psychological practice with transgender and gender nonconforming people*. Retrieved July 12, 2017, from http://www.apa.org/practice/guidelines/transgender.pdf

American Psychological Association, Task Force on Appropriate Therapeutic Responses to Sexual Orientation. (2009). *Report of the American Psychological Association Task Force on appropriate therapeutic responses to sexual orientation*. Retrieved June 9, 2017, from http://www. apa.org/pi/lgbc/publications/therapeutic-resp.html

Badgett, L., & Goldberg, N. (2009). *Best practices for asking questions about sexual orientation on surveys*. Los Angeles, CA: Williams Institute Sexual Minority Assessment Research Team.

Baum, K., Catalano, S., Rose, K., & Rand, M. (2009). *Stalking victimization in the United States* (NCJ 224527). Bureau of Justice Statistics Special Report. Washington, DC: U.S. Department of Justice. Retrieved May 9, 2014, from http://www.bjs.gov/index.cfm?ty=pbdetail&iid=365

Black, M. C., Basile, K. C., Breiding, M. J., Smith, S. G., Walters, M. L., Merrick, M. T., ... Stevens, M. R. (2011). *The National Intimate Partner and Sexual Violence Survey (NISVS): 2010 Summary Report.* Atlanta, GA: National Center for Injury Prevention and Control, Centers for Disease Control and Prevention.

Botuck, S., Berretty, P., Cho, S., Tax, C., Archer, M., & Cattaneo, L. (2009). *Understanding intimate partner stalking: Implications for offering victim services (NCJ Publication # 227220).* Washington, DC: National Institute of Justice, U.S. Department of Justice.

Breiding, M., Chen, J., & Black, M. (2014). *Intimate partner violence in the United States- 2010.* Atlanta, GA: National Center for Injury Prevention and Control, Centers for Disease Control and Prevention.

Buist, C. L., & Lenning, E. (2016). *Queer Criminology.* New York: Routledge.

Catalano, S. (2012). *Stalking victims in the United States - Revised.* Bureau of Justice Statistics Special Report (NCJ 224527). Retrieved July 12, 2014, from http://www.bjs.gov/content/pub/pdf/svus_rev.pdf

Centers for Disease Control and Prevention. (2014a). *The Relationship between bullying and suicide: What we know and what it means for schools.* Retrieved June 8, 2015 from http://www.cdc.gov/violenceprevention/pdf/bullying-suicide-translation-final-a.pdf

Finkelhorn, D. (2008). *Childhood victimization: Violence, crime, and abuse in the lives of young people.* New York, NY: Oxford University Press. doi:10.1093/acprof:oso/9780195342857.001.0001

Fisher, B. S., Reyns, B. W., & Sloan, J. J. (2016). *Introduction to victimology: Contemporary theory, research, and practice.* New York: Oxford Press.

Gladden, R. M., Vivolo-Kantor, A. M., Hamburger, M. E., & Lumpkin, C. D. (2014). *Bullying surveillance among youth: Uniform definitions for public health and recommended data elements, version 1.0.* Atlanta, GA: National Center for Injury Prevention and Control, Centers for Disease Control and Prevention and U.S. Department of Education. Retrieved May 15, 2017, from http://www.cdc.gov/violenceprevention/pdf/bullying-definitions-final-a.pdf

GLSEN, CiPHR, & CCRC. (2013). Out online: The experiences of lesbian, gay, bisexual and transgender youth on the Internet. New York: GLSEN.

GLSEN. (2017). *Gender Terminology, A discussion Guide*. Retrieved May 15, 2017 from https://www.glsen.org/sites/default/files/Gender%20 Terminology%20Guide.pdf

Greytak, E. A., Kosciw, J. G., Villenas, C., & Giga, N. M. (2016). *From Teasing to Torment: School Climate Revisited, A Survey of U.S. Secondary School Students and Teachers*. New York: GLSEN.

Hendricks, M. L., & Testa, R. J. (2012). A conceptual framework for clinical work with transgender and gender nonconforming clients. *Professional Psychology, Research and Practice*, *43*(5), 460–467. doi:10.1037/a0029597

Iyengar, S., & Kinder, D. (1987). *News that matters*. Chicago: Chicago University Press.

James, S. E., Herman, J. L., Rankin, S., Keisling, M., Mottet, L., & Anafi, M. (2016). *The Report of the 2015 U.S. Transgender Survey*. Washington, DC: National Center for Transgender Equality.

Jones, L. M., Mitchell, K., Wolak, J., & Finkelhor, D. (2013). Online harassment in context: Trends from three youth internet safety surveys (2000, 2005, 2010). *Psychology of Violence*, *3*(1), 53–69.

Jovonen, J., & Gross, E. (2008). Extending the school grounds? Bullying experiences in cyberspace. *The Journal of School Health*, *78*(9), 496–505. doi:10.1111/j.1746-1561.2008.00335.x PMID:18786042

Kann, L., Olsen, E., & McManus, T. (2016). Sexual identity, sex of sexual contacts, and health-related behaviors among students in grades 9–12 — United States and selected sites, 2015. *Morbidity and Mortality Weekly Report*, *65*(9), 1–202. doi:10.15585/mmwr.ss6509a1 PMID:27513843

King, G., Tomz, M., & Wittenberg, J. (2000). Making the most of statistical analysis: Improving interpretation and presentation. *American Journal of Political Science*, *44*(2), 347–361. doi:10.2307/2669316

Kosciw, J. G., Greytak, E. A., Bartkiewicz, M. J., Boesen, M. J., & Palmer, N. A. (2012). *The 2011 National School Climate Survey: The experiences of lesbian, gay, bisexual and transgender youth in our nation's schools*. New York: GLSEN.

Kosciw, J. G., Greytak, E. A., Giga, N. M., Villenas, C., & Danischewski, D. J. (2016). *The 2015 National School Climate Survey: The experiences of lesbian, gay, bisexual, transgender, and queer youth in our nation's schools*. New York: GLSEN.

Kosciw, J. G., Greytak, E. A., Palmer, N. A., & Boesen, M. J. (2014). *The 2013 National School Climate Survey: The experiences of lesbian, gay, bisexual and transgender youth in our nation's schools*. New York: GLSEN.

Kowalski, R., & Fedina, C. (2011). Cyber bullying in ADHA and Asperger syndrome populations. *Research in Autism Spectrum Disorders*, *5*(3), 1201–1208. doi:10.1016/j.rasd.2011.01.007

Lambda Legal. (2017). *Transgender Rights Toolkit, A Legal Guide for Trans People and Their Advocates*. Retrieved July 9, 2017, from https://www.lambdalegal.org/publications/trans-toolkit

Langevin, M., Bortnick, K., Hammer, T., & Wiebe, E. (1998). Teasing/bullying experienced by children who stutter: Toward development of a questionnaire. *Contemporary Issues in Communication Science and Disorders*, *25*, 12–24.

Leefeldt, E. (2016, May 10). *Victim of cyberbullying? Insurance might help*. Retrieved July 5, 2017, from http://www.cbsnews.com/news/victim-of-cyberbullying-insurance-might-help/

Lenhart, A., Smith, A., & Anderson, M. (2015). *Teens, technology and romantic relationships: Methods*. Retrieved December 8, 2016, from http://www.pewinternet.org/2015/10/01/teens-technology-and-romantic-relationships-methods/

Logan, T. (2010). *Research on partner stalking: Putting the pieces together*. Lexington, KY: University of Kentucky, Department of Behavioral Science & Center on Drug and Alcohol Research. Retrieved April 21, 2014, from http://www.nij.gov/topics/crime/intimate-partner-violence/stalking/documents/research-on-partner-stalking.pdf

Marcum, C. (2008). Identifying potential factors of adolescent online victimization for high school seniors. *International Journal of Cyber Criminology*, *2*(2), 346–367.

Minton, S. (2014). Homophobic bullying: Evidence-based suggestions for intervention programs. *Journal of Aggression, Conflict and Peace Research*, *6*(3), 164–173. doi:10.1108/JACPR-10-2013-0027

Mitchell, K., Finkelhor, D., & Wolak, J. (2003). The exposure of youth to unwanted sexual material on the internet: A national survey of risk, impact and prevention. *Youth & Society*, *34*(3), 3300–3358. doi:10.1177/0044118X02250123

Musu-Gillette, L., Zhang, A., Wang, K., Zhang, J., & Oudekerk, B. (2017). *Indicators of school crime and safety: 2016 (NCES 2017-064/NCJ 250650).* Washington, DC: U.S. Department of Justice.

National LGBTI Health Alliance. (2015). *'LGBTI' people and communities.* Retrieved May 22, 2017, from http://lgbtihealth.org.au/communities/#lesbian

National Telecommunications and Information Administration. (2014). *Exploring the digital nation: Embracing the mobile internet.* Retrieved February 19, 2015, from http://www.ntia.doc.gov/files/ntia/publications/exploring_the_digital_nation_embracing_the_mobile_internet_10162014.pdf

National White Collar Crime Center. (2013). *Cyberbullying, the legislative response, NW3C research brief.* Retrieved on May 11, 2017 from https://www.nw3c.org/docs/research/cyber-bullying.pdf?sfvrsn=6

NoBullying.com. (2015). *Ryan Halligan loses his life to taunts, rumors and cyber bullying*. Retrieved July 11, 2015, from https://nobullying.com/ryan-halligan

PACER National Bullying Prevention Center. (2017). *Questions answered.* Retrieved May 19, 2017, from http://www.pacer.org/bullying/

Rand, M., & Harrell, E. (2009). *National Crime Victimization Survey: Crime against people with disabilities, 2007 (No. NCJ 227814).* Washington, DC: Bureau of Justice Statistics.

Reisner, S. L., Greytak, E. A., Parsons, J. T., & Ybarra, M. L. (2015). Gender minority social stress in adolescence: Disparities in adolescent bullying and substance use by gender identity. *Journal of Sex Research*, *52*(3), 243–256. doi:10.1080/00224499.2014.886321 PMID:24742006

Reno, J. (1999). *1999 report on cyber stalking: A new challenge for law enforcement and industry*. Retrieved May 18, 2017, from https://www.justice.gov/archive/opa/pr/1999/September/421ag.htm

Reno, J. (2000). *Remarks of the Honorable Janet Reno, Attorney General of the United States to the National Association of Attorney Generals, January 10, 2000.* Retrieved May 19, 2017, from https://www.justice.gov/archive/ag/speeches/2000/011000naagfinalspeech.htm

Rose, C., Monda-Amaya, L., & Espelage, D. (2010). Bullying penetration and victimization in special education: A review of the literature. *Remedial and Special Education*, *32*(2), 114–130. doi:10.1177/0741932510361247

Saewyc, E. M., Bauer, G. R., Skay, C. L., Bearinger, L. H., Resnick, M. D., … Reis, E. (2004). Measuring sexual orientation in adolescent health surveys: Evaluation of eight school-based surveys. *Journal of Adolescent Health*, *35*, 345e.1–345e.16.

Schnell, P., & Garcia, M. M. (2013). Connecting the dots: The challenges of identifying and responding to stalking. *The Police Chief*, *80*(12), 62–64.

Sobey, D. (1994). *Violence and abuse in the lives of people with disabilities: The end of silent acceptance*. Baltimore, MD: Paul H. Brooks.

Spaulding, P. (2010, October 1). *Why did Tyler Clementi die?* Retrieved April 17, 2017, from http://edition.cnn.com/2010/OPINION/09/30/spaulding.rutgers.suicide/

Substance Abuse and Mental Health Services Administration. (2014). *A Practitioner's Resource Guide: Helping Families to Support Their LGBT Children. HHS Publication No. PEP14-LGBTKIDS*. Rockville, MD: Substance Abuse and Mental Health Services Administration.

Substance Abuse and Mental Health Services Administration. (2015). *Ending Conversion Therapy: Supporting and Affirming LGBTQ Youth. HHS Publication No. (SMA) 15-4928*. Rockville, MD: Substance Abuse and Mental Health Services Administration.

Sullivan, P., & Knutson, J. (2000). Maltreatment and disabilities: A population-based epidemiological study. *Child Abuse & Neglect*, *24*(10), 1257–1273. doi:10.1016/S0145-2134(00)00190-3 PMID:11075694

Tjaden, P., & Thoennes, N. (1998). *Stalking in America: Findings from the National Violence Against Women Survey (Research in Brief NCJ 169592)*. Washington, DC: National Institute of Justice and Centers for Disease Control and Prevention.

Tomz, M., Wittenberg, J., & King, G. (2003). CLARIFY: Software for interpreting and presenting statistical results [Software]. Available from https://gking.harvard.edu/publications/clarify-software-interpreting-and-presenting-statistical-results

United Nations Children's Fund. (2014). *Hidden in plain sight: A statistical analysis of violence against children*. New York: UNICEF.

United States Department of Education. (2013). *Student report of bullying and cyber-bullying: Results from the 2011 school crime supplement to the National Crime Victimization Survey*. Retrieved March 23, 2014, from http://nces.ed.gov/pubs2013/2013329.pdf

United States Department of Health & Human Services. (2015). What is cyberbullying? Retrieved January 5, 2015, from http://www.stopbullying.gov/

United States Department of Health & Human Services. (2017). *Prevent bullying*. Retrieved May 12, 2017, from https://www.stopbullying.gov/prevention/index.html

Wolak, J., Mitchell, K., & Finkelhor, D. (2007). Unwanted and wanted exposure to online pornography in a national sample of youth internet users. *Pediatrics*, *119*(2), 247–257. doi:10.1542/peds.2006-1891 PMID:17272613

Wong, C. (2012, April 18). *Heartbreak details in gay Iowa teen's suicide emerge*. Retrieved June 25, 2017, from http://www.huffingtonpost.com/2012/04/18/gay-iowa-teen-death-details_n_1434899.html

Zweig, J. M., Dank, M., Lachman, P., & Yahne, J. (2013). *Technology, teen dating violence and abuse, and Bullying*. Urban Institute's Justice Policy Center, Document No. 243296. Retrieved January 9, 2015, from https://www.ncjrs.gov/pdffiles1/nij/grants/243296.pdf

Chapter 3
Legislative Response to Cyber Aggression:
Federal and State-Local Policy Reform

ABSTRACT

This chapter presents the federal and state-local legislative response to cyber aggression: stalking, harassment, and bullying. Along with other federal efforts, the federal Violence Against Women Act and its reauthorizations is identified as a cornerstone law in protecting the public on stalking and harassment. State-local laws have advanced in scope; yet, there are laggard states not yet entirely on board in passing legislation aligned with the advancement of technology used in cyber aggression. All three branches of government to some extent have had a voice in today's cyber policy. Judicial court cases have shaped policy decisions and several key cases are presented.

INTRODUCTION

Cyber aggression takes on varying forms including stalking, harassment, bullying, and nonconsensual pornography. See Table 1 of Chapter 1. These are behaviors restricted by laws and court rulings. Federal and state regulatory and administrative legislation fighting these dysfunctional behaviors have been incremental in the making. On the federal level, the 1994 Violence against Women Act (Public Law 103-322) is a cornerstone law to supporting stalking victims such as women and children. It authorized grants to states

DOI: 10.4018/978-1-5225-5285-7.ch003

and tribal government to fight the aggression on the domestic level. With cyberbullying, there is no chief federal law that governs over the behavior. Instead, the federal government has devolved authority to states and school districts, and judicial decisions have had an impact on school discipline policy. In the fight against cyber aggression, states know the protective needs of their local best, but not all are on board. While the majority has some type of law protecting residents from "physical" aggression crimes, not all have updated to include the "cyber" of aggression. This chapter discusses government action or inaction on enacting regulatory and administrative cyber aggression laws on stalking, harassment, bullying, and nonconsensual pornography.

BACKGROUND

With nearly half (49%) of the world online (Pew Research Center, 2017) more people are susceptible to cyber aggression as criminals use technology to cyber stalk, harass, and bully. This is not surprising since obtaining electronics for criminal intent is easy. Most anyone can purchase a computer, multifunctional cell phone device and supporting software; and access to the World Wide Web is becoming less costly and free in some places. The traditional "physical" stalking, harassment, and bullying behaviors now have a digital counterpart. The U.S. Department of Health and Human Services (2015) reports cyber aggression as an "emergent concern," and not limited to "sending threatening texts, posting or distributing defamatory or harassing messages, and uploading or distributing hateful or demeaning images or videos intended to harm another" (USDHHS, 2015). Concerned about cybercrime, some Americans are taking counter measures to confront the unfriendly digital climate. A 2013 Pew Internet & American Life Project survey found as many as 55% of respondents reported avoiding online observation by people, employers, government, organizations, and other; while 86% of adult Internet users have taken measures to promote anonymity, privacy, and security online. To avoid surveillance, online safety behaviors range from masking personal information, clearing search histories, to using a public computer instead of personal home computer (Pew Internet & American Life Project, 2013).

The definition of stalking varies from state to state. Generally, it is "a course of conduct directed at a specific person that causes actual fear or would cause a reasonable person to feel fear" (U.S. Department of Justice Office on

Violence against Women [USDOJOVW], 2017, p. 6). The cyber counterpart is the use of the Internet, email or other electronic communications to stalk, and generally refers to a pattern of threatening or malicious behaviors. It may be considered a most dangerous type of Internet harassment. The sanctions range from misdemeanors to felonies (National Conference of State Legislatures [NCSL], 2013a). This crime is under federal, state, District of Columbia and U.S. territories law (USDOJ, 2012). Cyber-harassment differs from cyberstalking since it may not involve a credible threat, and usually pertains to threatening or harassing email messages, instant messages, or to blog entries or websites dedicated solely to tormenting an individual (NCSL, 2013a), reference Table 1 of Chapter 1 on definitions. Not all states are keeping up with provisions that cover advanced technology methods being used by perpetrators for cybercrimes. Although federal laws (Violence against Women Act of 2005 & 2013) extended protective regulation to cover both interstate and intrastate cyberstalking, many states have not caught up with the newer forms of technology, such as Global Positioning Systems surveillance and videotaping. Some state laws use "broader language" to cover the many types of evolving stalking methods (physical and electronic) used by perpetrators, whereas others have separate laws (USDOJOVW, 2017; NCSL, 2013a). Some states approach cyber-harassment by including language addressing electronic communications in general harassment statutes, while others have created stand-alone cyber-harassment laws (NCSL, 2013a). Chapter 5 brings an empirical insight on why some states have taken the basic steps of updating existing stalking and harassment laws to now include electronic communication technology, and others not.

With cyberbullying, much like physical bullying, girls are more likely to be victims than boys. Popular methods are Internet instant messaging, chat rooms, e-mails and posted website messages (NCSL, 2011). There are no federal laws "directly addressing" bullying per se, as the responsibility has been devolved to the local government for policy making. Defining cyberbullying has not been clear cut for the states. While some states differentiate bullying from harassment, others say the two cannot be differentiated. The cyberbullying language in some state laws is the same for cyber-harassment. "In some cases, bullying overlaps with discriminatory harassment when it is based on race, national origin, color, sex, age, disability, or religion" (Antibullying Institute, 2017, p. 1).

As described in Chapter 1, cyber revenge pornography is a form of aggression facing minors and adults. Some refer to the behavior as sexual cyber-harassment, involuntary or non-consensual pornography, whereas doxing refers to the posting of personal information about another in an act of revenge. Cyber pornography involves "the dissemination or posting sexually explicit media without the consent of the individual in the media, particularly where the intent is to shame, humiliate, and frighten the person or otherwise cause them harm" (Lonardo, Martland & White, 2016, p. 80). A shorter definition by the State of Arkansas's Act 304 is the "unlawful distribution of sexual images or recording." Both the federal government and states are assuming a strict stance on this problem. All states have passed some type of legislation governing child pornography (Cyberbullying Research Center, 2015c).

GOVERNMENT RESPONSE TO CYBER AGGRESSION

In effort to protect the public from cyber aggression, federal laws have been passed and several are in response to some violent crisis event and/ or vulnerable group most in need of law. The federal government has made a commitment to assist the states on domestic crime with funding. Cyberstalking and harassments policies are largely aimed to protect females; cyberbullying is primarily geared to protect children and youth. Updating federal anti-discrimination and harassment laws to protect the lesbian, gay, bisexual, and transgender population has progressed.

The federal and state-local response to cyber aggression is primarily regulatory and administrative law. Lowi (1972) classifies public policy as falling into four major categories: distributive, re-distributive, regulatory, and constituent policy. Policy concerning the criminalization of cyberstalking, harassment and bullying activities can be conceptualized as regulatory and administrative policy. Specifically, these protective-type regulatory policies are designed to protect society by prohibiting harmful conditions (Ripley and Franklin, 1980, pp. 20, 24), and considered "only one of several ways governments seek to control society and individual conduct" (Lowi, 1972, p. 299). The political relationships in policymaking of regulatory policy are largely Congressional committees and sub-committees, the full House and Senate, executive agencies, and trade associations (Ripley and Franklin, 1980, p. 22). On the other hand, administrative reform policy (McNeal, Tolbert, Mossberger & Dotterweich, 2003), unlike regulatory policy does not involve the direct and coercive use of government power over citizens (Lowi, 1979). Administrative policy tells us "how" policy/ programs are to be carried out, and are technical rather than value-laden and non-salient with the public.

Major players tend to be administrative officials even though state legislators play a role through budget processes or legislative oversight and professional networks (McNeal et al., 2003). In all, protecting against cyber aggression requires not only federal but state-local regulatory and administrative laws.

Violence Against Women Act as a Federal Cornerstone Law

Current federal laws fighting cyberstalking and cyber-harassment are largely built on early crime laws aimed at public safety. Initially, Congress enacted regulations protecting the public from violence such as (physical) stalking and harassment. The "cyber" counterpart was added to law by subsequent provisions. Prior to the 1970s, women were invisible victims of violent crimes, such as battering, harassment, stalking and policy centered on family domestic violence. For example, considered a domestic family issue, in 1984 the Family Violence Prevention and Services Act (Public Law 98-457) was passed assisting states to prevent family violence and to provide shelters and services to those victimized (Congressional Research Service, 2014). Modern day cyberstalking laws protecting women are rooted in these older concepts, and as family violence took on a modern-day social view, the laws were modernized. It was in the 1990s that society began to view stalking as a criminal conduct as falling within the jurisdiction of the federal and state criminal justice systems. Considered a legislative milestone in dealing with gender based violence against women, the 1994 Violence against Women Act often referred to as VAWA was enacted by Congress as Title IV of the Violent Crime Control and Law Enforcement Act of 1994 (Public Law 103-322). The 1994 VAWA gave state, local, and tribal law enforcement jurisdictions grant opportunities for the investigation and prosecution of violent crimes against women---domestic violence, sexual assault, dating violence, stalking, and other (USDOJ, 2016). "VAWA articulates the Congress's commitment to effective strategies for preventing and responding to domestic and sexual violence, holding offenders accountable, and ensuring safety, autonomy, and justice for victims" (USDOJOVW, 2015b, p. ix).

Much like the Civil Rights Act of 1964 (Public Law 88-352) with all its amendments improving the gesture of the law over time, the VAWA has expanded in scope now giving many a biased/ vulnerable group their law enforcement attention by way of federally funded programs on anti-stalking and harassment. The VAWA was reauthorized in 2000 and 2005 with broad bipartisan support, and again in 2013 with "some adversity" (Congressional Research Service, 2015). In both 2000 and 2005 the scope of protection and services for victims were expanded. Specifically, the Victims of Trafficking and Violence Protection Act of 2000 (Public Law 106-386) reauthorized the 1994 VAWA and its grant monies for policy implementation, in particular to support law enforcement actions that encouraged arrest policies. In 2000, other VAWA monies went to states for their domestic violence and sexual assault coalitions; and to Indian tribal governments to cover rural domestic violence and child abuse. The law improved the access to protection of battered immigrant women. In 2005, Congress reauthorized appropriations for VAWA through the Violence against Women and Department of Justice Reauthorization Act (Public Law 109-162) that modernized the goals of VAWA. Title I of P.L. 109-162 enhances judicial and law enforcement tools to fight violence against women. Title II improves services for victims of domestic violence, sexual assault, and stalking. Title V strengthens the healthcare system's response to aggressive crime. Among funding appropriations was a new program developed for cultural and linguistic services for victims because of today's changing cultural landscape; monies went to servicing children and women age 50 and over who are victim vulnerable; and higher education campus needs. Prioritized for monies was the underserved and rural American. The VAWA 2005 now gives the Attorney General power to authorize grants to Indian tribal governments and those organizations aimed at improving the capacity of Indian Tribal Government service to native women (Public Law 109-162, 2005) as these women have been victims to a high incidence of aggression. Since stalking methods have become more advanced through new technology (Internet, global positioning systems, etc.), the 2005 Act introduced "interstate cyberstalking" (USDOJOVW, 2017, p. 7). More recently in 2013, Congress reauthorized the VAWA through the Violence against Women Reauthorization Act of 2013 (Public Law 113-4), which expanded the definition of stalking to now include "intrastate cyberstalking" crimes. 2013 also reauthorized many of the existing grant programs for states and tribal governments from FY2014 through FY2018. The 2013 law further improved upon 2005 service access for many vulnerable groups historically discriminated (Public Law 113-4, 2013). In all, the federal government has passed many a law, but does not act alone in prohibiting cyberstalking and harassment. It has committed to provide national monetary aide through grants to lower levels of government, and makes federal agency assistance possible (falls under the executive branch of U.S. government) e.g. the U.S. Department of Justice and U.S. Department of Education.

Grant Funding to Fight Cyber Aggression

In the U.S., all 3 levels of government may be involved in putting a single piece of legislation into effect (Dye, 2008). The U.S. has created federal agencies to apply the principles of legislation, monitor policy implementation and evaluate its effectiveness; federal agencies largely collaborate with state-local agency bureaucrats in policy implementation and money is involved. In the case of the Violence against Women Act, a vertical implementation structure is used involving federal, state and local institutions and a host of policy actors. In 1995, the VAWA charged the Department of Justice to administratively create its Office on Violence against Women (OVW) to oversee VAWA grants, as grants are the primary method in funding the VAWA policy programs. Policy implementation is made difficult without funding. The OVW awarded greater than $6 billion in grants and cooperatives to state-local and tribal governments, non-profits and universities (Congressional Research Service, 2015, p. 4). The U.S. Department of Justice Office on Violence against Women and the National Center for Victims of Crime partnered in year 2000 to create a one-stop resource center to effectively interface with stalking in the U.S. This Stalking Resource Center provides government, practitioners, and victims with tools to improve the response of the criminal justice system, enhance victim safety, hold offenders accountable, give training on multiple related topics, etc. (Stalking Resource Center, 2017). The National Center for Victims of Crime has provided advocacy for victims of crime across the U.S. since 1985.

The grant programs are an important policy implementation tool for VAWA giving victims protection and concurrent services to fight aggression. Since 1994 through 2016, VAWA provisions were added authorizing multiple types of grants (formula and discretionary) to state governments, enforcement agencies, coalitions, including the actual victims of crime. Reported in 2016, the VAWA Office on Violence against Women administered 15 discretionary grants and 4 formula grant programs giving monies to institutions on domestic and sexual violence (p. vii). These grants cover many vulnerable women groups, including "culturally specific" underserved groups as with the Culturally Specific Services Program (CSSP). The 2013 VAWA reauthorization made "culturally specific" to mean racial and ethnic minority groups. The CSSP is a community program aiding racial/ ethnic minorities, and its subsequent 2014 expansion added victims of deaf or hard of hearing, underserved religious/ ethnic groups, and aiding the LGBT (lesbian, gay, bisexual, and transgender) populations (USDOJOVW, 2016, p. 130). As campus aggression has increased, VAWA grants also apply to higher education institutions. The Grants to Reduce Violent Crimes against Women on Campus Program aims at improving campus security, and the prevention and prosecution of aggression. The Department of Justice reports that 92 campus grants were awarded to 31 states for 2012-2015 (USDOJOVW, December 2016, p. 3-4).

A grant program awarded in large number is the Services, Training, Officers, Prosecutors Violence against Women Formula Grant Program [STOP] (USDOJ, 2015). The STOP grant formula is based on population and awarded to states and territories. The DOJ reports $269,532,798 awarded to states and territories during 2011 and 2012, with sub awards of $255,203,456 distributed to victim service agencies and organizations, coalitions, law enforcement, prosecutor offices, courts and other sub grantees (USDOJOVW, 2015b, p. x). Sub grantees receiving STOP funding have used the monies in various ways. The Commonwealth of Virginia, for example, has enhanced its Domestic Violence Intervention Project, educating its STOP stakeholders (including law enforcement and prosecutors) on the new technologies being used by stalkers. In Pennsylvania, the Helping All Victims in Need Project trains law enforcement "how" to build felony stalking charges and work with other municipalities. Ohio's Stalking Investigator initiative funds investigator liaison with multiple jurisdictions, to conduct surveillance on suspects, and funds the expertise to compile profiles of suspects to develop a big picture of the case (USDOJOVW, 2015b, p. 63-64). The total number of STOP sub grantee awards distributed across the states in 2011 totaled 3,338 led by Michigan at 370, followed by Ohio at 255, New York at 130, Minnesota 121, Maryland 119, Texas 116, and Utah at 101 awards (USDOJOVW, 2015b, p. 118). The U.S. Department of Justice (2015) reports domestic violence against American Indian and Alaskan Native women on reservations is high at "10 times the national average" (p. 3). In response to need, the 2013 VAWA broadened the responsibility of states distributing STOP funding to include tribal governments as potential recipients of monies. In addition, the Grants to Encourage Arrest Policies and Enforcement of Protection Orders program is also available, encouraging state-local government and the tribal government and its courts to take domestic violence / stalking as a "serious" crime encouraging arrest policies and policy enforcement (Congressional Research Service, 2015, p. 18-21).

Aside from the VAWA, the federal government has made other monetary commitment to help the states fight domestic crime with funding for prevention, prosecution, safety and housing, etc. The federal Omnibus Crime Control and Safe Streets Act of 1968 (Public Law 90-351) devolved much responsibility for citizen protection to the state-local governments. Here, Congress pledged to provide them with national monetary assistance through grants to prevent, detect and apprehend criminals, including the crime against women. This 1968 public law also prohibited the invasion of privacy done through wiretapping and electronic surveillance by criminals, thus laying a foundation for future laws prohibiting perpetrator surveillance as done in cyberstalking and cyber-harassment. In 1994, the Violent Crime Control and Law Enforcement Act (Public Law 103-322) authorized monies for transitional housing for female victims of domestic violence, sexual assault and stalking. In 2009, the American Recovery and Reinvestment Act (Public Law 111-5) authorized grant monies for state and local law enforcement activities connected to the Violence against Women Prevention and Prosecution Programs including monies for the victim housing assistance grants, previously authorized by the 1994 law.

Some may not consider the 1994 Violence against Women Act as the only cornerstone federal law on cyberstalking. Alongside the VAWA and its modernizing reauthorizations, 3 federal codes are considered in fighting cyberstalking---18 USC § 2261A, 18 U.S. Code § 875, and U.S. Code 47 § 223. The U.S. Code Title 18, Domestic Violence and Stalking §2261A added the use of "any interactive computer service or electronic communication service or electronic communication system of interstate commerce" to existing federal laws governing the physical stalking activities, and expanded the scope of stalking crimes to include the "intent" and "effects" of stalking on victims (18 USC § 2261A, 2012; University of North Carolina at Chapel Hill [UNC], 2017a; UNC, 2017b). The 18 U.S. Code § 875 (2012) added the cyberstalking penalties, and the U.S. Code 47 § 223 (2012) outlined the prohibited acts of obscene or harassing telephone calls (wire or radio telecommunications) for interstate or foreign communications and for the District of Columbia; covered the legal enforcement of stalking laws; and covered the defense to prosecution (UNC, 2017a; UNC, 2017b).

State-Local Cyberstalking Laws

States have picked up where federal government has left off in adopting cyberstalking laws to protect and punish. State laws are non-uniform. They vary in the modernization of old offline stalking laws to cyberstalking laws. In Chapter 5, we analyze over time why some states modify existing stalking or harassment legislation to include electronic communication and why others are laggard using regression and controlling for political constraints, interest group strength, state wealth, and state need or demand for updated laws. While some states are laggard others are leaders in passing cyberstalking laws. California is considered a leader having enacted a broad scope of administrative and regulatory laws protecting Californians on cyberstalking. This may be expected as California had the highest number count of victim complaints on Internet crime at 39,547 as submitted to the Federal Bureau of Investigation's Internet Crime Complaint Center for 2016 (FBI, 2017, p 19), see Table 1 *Victim Complaints about Internet Crime to FBI, 2016.* Historically, California was the first state to codify stalking as a crime in 1990, largely a result of homicide cases involving women victimized by stalking. In 1992 and 1993, an additional 45 states followed to enact laws criminalizing stalking (National Institute of Justice, 1996, pp. A1-A11). Today, California has both criminal and civil laws governing cyberstalking and cyber-harassment. They have expanded the scope of harassment, which can be a single "intentionally harassing contact" of threat or can be "obscene" language. Among California laws, personal information cannot be electronically disseminated to harass or cause emotional distress; the posting of pictures on the Internet of victims or revenge porn is illegal without permission; penalty is more severe with recidivism or minor age victims (Privacy Rights Clearinghouse, 2016). Theoretically, states learn about "good" policy ideas and "poor" ones from each other, and California has been observed by many for being a leader and innovative across different policy domains. Laggard states can save policy money by learning from the mistakes and practices of California before adopting their own laws.

Are these state laws adequate to protect victims? According to Goodno (2007), "state statutes that might be used to prosecute cyberstalking do not have clear and equal standards" and are inadequate to deal with the distinct issues of cyberstalking (p. 140, 156). The U.S. Department of Justice (2012) reports that among the differences, states vary relative to the level of fear and distress necessary for the stalking behavior to be elevated to the level of

Table 1. Victim complaints about Internet crime to FBI, 2016

States With Highest # of Complaints	Count	States With Lowest # of Complaints	Count
California	39, 547	North Dakota	350
Texas	21,441	South Dakota	376
Florida	21,068	Wyoming	432
New York	16,426	Vermont	440
Illinois	9,177	Rhode Island	663
Maryland	8,361	Delaware	703
Pennsylvania	8,265	Montana	744
Virginia	8,068	Maine	770
Ohio	7,052	Nebraska	1,028
Washington	6,874	Hawaii	1,055

Note: The numbers in Table 1 are not a rank ordering since they do not consider the multiple variables involved with each crime case; they are based on the total number of victim complaints in each state, and territory including D.C.; count is aggregate for all Internet crimes but do include harassment/ threats of violence, social media, crimes against children. Source: Federal Bureau of Investigation Internet Crime Complaint Center. (2017), 2016 Internet Crime Report, p. 19.

criminal and perpetrator intent. "Actual fear" typically requires testimony by the victim relative to lifestyle changes because of stalking. While some laws require actual fear by the victim, other laws have a "reasonable person standard" where the perpetrator behavior would cause a reasonable person to experience fear (USDOJ, 2012, p. 3). In their evaluation of cyberstalking laws across the states, Goodno (2007) recommends effective cyberstalking statutes to "criminalize conduct that either puts a 'reasonable person' in fear of bodily harm or causes severe emotional distress" and statutes need to directly target the situations where the cyber-stalker gets innocent third parties to do the harassing and stalking for them (p 133, 156). Goodno reports, using the reasonable person standard provides the "most successful way to prosecute cyberstalking" because it does not require physical proximity as with offline stalking; it addresses many of the issues resulting from a credible threat requirement; it does not require an unequivocal threat being sent to the victim; does not require the victim to prove that the cyber-stalker had the ability to follow through on the threat. Goodno (2007) considers the reasonable person standard as the appropriate standard to be used in cyberstalking laws (p. 139-140).

Aside from a variation on level of fear and distress, crime can vary according to general or specific intent. Most crimes require "general intent" meaning that the accused did what the law prohibits. Whereas, "specific intent" crimes typically require that the accused did what the law prohibits and intended to accomplish the precise act the law prohibits (Black's Law Dictionary, 1990, p. 810). States also vary on the level of severity of the crime committed. According to the Stalking Resource Center (2015), "less than 1/3 of states classify stalking as a felony upon first offense. More than 1/2 of states classify stalking as a felony upon second or subsequent offense or when the crime involves aggravating factors" such as victim age under 16 years and victimization more than once (p 1).

Cyber Pornography Laws

Revenge pornography is a form of harassment/bullying. Cyber pornography is governed by a collection of laws that regulate perpetrator use of electronic means to transmit sensitive images of child porn. In 1967, the U.S. Congress called obscenity and pornography a "national concern" asking for a coordination of all levels of government on the matter, and created an advisory Commission on Obscenity and Pornography. Made up of multidisciplinary specialists, they researched the causal relationship of pornography to antisocial behavior to give "advisable action" to Congress (Public Law 90-100, 1967) and established a scientific understanding on this type of aggression. It was later that advanced technology use in pornography, such as the computer had made the government decision making agenda. Responding to the visual possibilities that could be created using the Internet at that time, the Child Protection Act of 1984 (Public Law 98-292) updated an older law, the Protection of Children against Sexual Exploitation Act of 1977 (Public Law 95-225) modernizing the language of pornographic sensitive images that previously meant "obscene visual or print medium" to the newer language of "visual depiction" of pornographic material. The 1984 Child Protection Act also strengthened the criminal and civil forfeiture of the property used in the pornography crime, and reinforced the customs laws on judicial forfeiture of property. Another law advancing the regulations against cyber revenge pornography is the 1988 Child Protection and Obscenity Enforcement Act of 1988 (Public Law 100-690). This 1988 law made it illegal to use a computer to transport, receive or possess "with intent to distribute" visual depiction of minors either interstate or in foreign commerce. Later, the 1998 Protection of Children from Sexual Predators Act, Title II, expanded the "jurisdictional base for prosecution of production of child pornography" to include the use of the computer in receiving and handling of materials for a visual depiction of a child (Public Law 105-314, 1998). In 1996, the Child Pornography Prevention Act (H.R. 4123, 1995-1996) closed loopholes from previous laws and again updated the capabilities of advanced technology on pornography. Since newer technology had already evolved with capability to create and further modify images of children making them "virtually indistinguishable" from real images, the 1996 law also broadened the definition of a "visual depiction." Some defendants claim the newer definition violated their First Amendment rights and challenged it in court.

In addition to the collection of federal laws that govern over child pornography, the states are passing their own sexting laws. As defined in Chapter 1, sexting is the sending of a naked or partially naked picture of oneself to another person using an advanced technology medium typically done through text messaging. In 2013, about 20 states had some type of law covering sexting by minors, 26 states had laws on cyber revenge pornography, and 18 states had no law. Penalty laws vary. For example, in Utah, the first offense for minors who are involved in sexting is considered a misdemeanor, but considered a felony for all other offenses. In Rhode Island, the minor disseminating a sexually sensitive image of self to others is committing a "status offense" and sent to family court (Cyberbullying Research Center, 2015c). Adults, however, can receive a stricter penalty when it comes to cyber pornography as seen in Illinois state law. Illinois violators get court determined forfeiture of property (such as computer or even profits) and forfeiture of any "contractual right of any kind affording a source of influence over any enterprise" that is related to child pornography, aggravated child pornography, or "non-consensual dissemination of private sexual images" (State of Illinois, 2015).

Florida is a leader, having passed a range of sexual cyber-harassment and revenge pornography laws with implementation beginning in 2015. This may be expected (analyzed in Chapter 5), as Florida followed California and Texas at a count of 21,068 in the number of Internet crime victim complaints submitted to the FBI for 2016 (FBI, 2017), see Table 1. Florida defines "image" of a sexually explicit image to include but not limited to any photograph, picture, motion picture, film, video, or representation. Their penalties vary and include misdemeanor of the first degree and felony of the third degree for the person who has one prior conviction and commits a second or subsequent offense (Florida Legislature, 2015). In cases of multiple images transmitted within 24 hours it is considered a "single" offense. Florida penalty allows for counseling or some other "informal" type of sanction, and persons can seek civil remedies. Their law addresses both the sending and receiving of sexting for those under age 18 years. Minors receiving images without request and not having distributed the image(s) and having reported the incident to authority are not considered to have committed a sexting offense (Cyberbullying Research Center, 2015c). Florida authorizes law enforcement to arrest without warrant anyone that he or she has probable cause to believe has committed sexual cyber-harassment; authorizes a search warrant issued in "specified" instances; specifies the circumstances of offense, and more (Florida Legislature, 2015). Chapter 5 empirically explores why some states have a higher legal penalty for committing a crime of revenge porn while others do not.

Cyberbullying Laws

Although the federal government has been active in passing cyberstalking laws to protect women, it has devolved responsibility on cyberbullying and accompanying child harassment onto the states and its school districts. There is no cornerstone federal law for cyberbullying as seen with cyberstalking. The federal government has however governed over schools receiving federal funds by requiring them to assume a serious anti-bullying agenda. Recently, the 2017 Congress the Tyler Clementi Higher Education Anti-Harassment Act of 2017, not yet law, was introduced as House bill 2151 and as Senate bill 954 aimed at the higher education setting. This bill targets harassment in higher education that is associated with a student's (actual or perceived) race, color, national origin, sex, sexual orientation, gender identification, disability, or religion. It defines electronic communication used in cyber harassment broadly so as to reflect newer cyber methods: "any transfer of signs, signals, writing, images, sounds, or data of any nature transmitted in whole or in part by a wire, radio, electromagnetic, photo electronic, or photo optical system" (House of Representatives 2151, June 2017). As discussed in Chapter 2, based on its legislative history, the Tyler Clementi Higher Education Anti-Harassment Act of 2017 is not expected to become law. Currently in Congressional committee, the Safe Schools Improvement Act of 2017 targets student bullying and harassment (House of Representatives 1957, 2017). It would amend the Elementary and Secondary Education Act of 1965, which is aimed "to strengthen and improve educational quality and educational opportunities in the Nation's elementary and secondary schools" (Public Law 89-10, p. 27). As a federal incentive to get local level government to move on anti-bullying and harassment of youth, the Safe Schools Improvement

Act of 2017 would require those districts receiving federal monies to adopt codes regulating school bullying and harassment of students (Human Rights Campaign, 2017; H.R. 1957). Schools receiving federal funding are expected to resolve bullying and harassment complaints, cyber and otherwise. Several federal agencies are available to aid them; the U.S. Department of Education Office for Civil Rights and the U.S. Department of Justice Civil Rights Division are available to assist (Antibullying Institute, 2017, p. 1).

Basically, cyberbullying of youth is considered a local policy matter best resolved by the locale more in tune with the regional needs. To keep up with digital advances used by bullies, some states have been innovative through adopting laws in response to their devolved authority, whereas other states remain laggard. States are experimenting with legal strategies. Many a state cyberbullying law overlaps with its harassment laws. Since perpetrator harassment can spill over into cyberstalking, some states reflect this by covering many of the types of cyber aggression in a single law, while others are more specific to differences in crime. For example, Minnesota and Nebraska's statutes on stalking and harassment can also be applied to prosecution of cyberbullying. Kentucky considers cyberbullying between students as part of its concept of cyber-harassment, punishable as a class B felony. New Mexico defines electronic bullying to include but not limited to "hazing, harassment, intimidation or menacing acts of a student which may, but need not be based on the student's race, color, sex, ethnicity, national origin, religion, disability, age or sexual orientation" (NCSL, July 2010; January 2013b).

As with cyberstalking, cyberbullying laws passed by states are largely regulatory and administrative. In addition to enacting anti-bullying regulations governing the entire state, states are also passing administrative laws devolving much authority to the local school districts to implement the state law, to design procedures on enforcement, and to create their own local policies tailored to their specific needs. New Hampshire for example devolves authority to the school board of every school district, while in Idaho power is also given to school officials to penalize cyber-harassment. Oregon empowers community stakeholders in effort to push a local "participatory" policymaking idea with "parents, school employees, volunteers and community representatives" to model their laws (NCSL, January 2013b). Vermont promotes student participation in advisory adding a secondary student to their Advisory Council on Harassment, Hazing, and Bullying (NCSL, January 2013). Some states have codified federal standards into statute, such as federal case laws allowing "schools to discipline students for off-campus behavior that results in a substantial disruption of the learning environment at school"

(Cyberbullying Research Center, 2016, p. 1). Florida's 2008 Jeffery Johnston Stand Up for All Students Act requires that if school networks are being used by students, then Internet bullying means "on or off" the school zone. In addition to school zone, California extends the scope of cyberbullying laws to include school personnel (NCSL, 2013b). In all, the state government and local school districts (government units) have assumed an intergovernmental policy relationship with their anti-cyberbullying campaign.

Theoretically, policies tend to remain stable for extended periods of time until new information or events take place drawing attention to a problem and elevating it onto political agenda (Baumgartner & Jones, 1993; Cobb & Elder, 1971). This is consistent with the finding that legislative action on school bullying occurred after a focusing event such as the 1999 Columbine (Colorado, USA) school shootings in 1999. As a result of this 1999 crisis event, Georgia was a pioneer in adopting anti-bullying laws, including a law requiring its schools to implement "character education programs" that specifically addressed bullying prevention. Forty-six states followed Georgia from 1999-2000 enacting more than 120 separate bills either introducing or amending bullying statutes in their education or criminal codes. The four states that were laggard in following Georgia's lead were Hawaii, Michigan, Montana, and South Dakota who did not pass these laws (U.S. Dept. of Education, 2011). Except for Michigan (historically an early adopter of policy across many different policy domains), Hawaii, Montana and South Dakota may have a lesser need to adopt Internet crime laws (see Table 1). Just as Georgia pioneered in anti-bullying laws, Utah in 2001 became the first to include language for cyberbullying. More legislative action on the "cyber" of bullying did not happen until 2007 (Cyberbullying Research Center, 2015a) and state momentum picked up around 2010 where 21 new bills were passed, and 8 bills were signed into law through April 30, 2011 (U.S. Dept. of Education, 2011). In contrast to Utah being an early adopter, Montana has been laggard in passing cyberbullying laws. In 2015, Montana became the last state to adopt policy and practically speaking, it was strictly symbolic. The law provided a definition of bullying but did not mandate any actions to be taken. Montana is not a state with a high need for cyber policy given that its Internet crime rate is amongst the lowest reported to the FBI. In contrast to the majority of states, the FBI (2017) shows Montana having low reported Internet crime, based on a count of 744 victim complaints in 2016 (see Table 1). This suggests an association between the occurrence of Internet crime and the need for law; empirically explored in Chapter 5.

Today, according to the Cyberbullying Research Center (2017), 49 states require schools to have a policy governing bullying, 16 states have laws that include off-campus behaviors, and 44 states have criminal sanctions for cyberbullying or electronic harassment (2017). Those 6 states lacking criminal sanctions are Maine, Minnesota, Nebraska, New Hampshire, New Mexico, and Wyoming (Cyberbullying Research Center, 2017). Except for Minnesota, the 5 others had a lower number of reported Internet crime for 2016, which may or may not be a reason for not passing criminal sanctions, see Table 1 (FBI, 2017). The fight against bullying is not limited to state government. As a leader, the capital of the U.S., District of Columbia, adopted the Youth Bullying Prevention Act of 2012, which is an administrative law framing "how" bullying prevention efforts may be modeled by stakeholders citywide. It uses "a public health framework with three levels of prevention practices and strategies." The first level of prevention practices and strategies is on the persons in need of prevention action; the second is on persons at risk and the places of risk for this victimization; and third is the strategic responses to the incidents (Urban Institute Justice Policy Center, 2013, p. 3).

State bullying laws can also be shaped by judicial decisions on school discipline or the lack of discipline to bullying and harassment. In *Zeno v. Pine Plains Central School District* (2012) the school failed to adequately protect a student from harassment. The plaintiff, a high school student, received a final award of $1 million in damages because his school failed to protect him from harassment (on and off school property) for nearly 3-and-a-half years. The student was subjected to "verbal racial attacks" and physical attacks. Although the school disciplined students for the bullying and suspended one student for 45 days, the school district had not taken "definitive action" ensuring that the aggression had stopped. Zeno brought action "contending that the District was deliberately indifferent to his harassment." A jury found the District liable in violating Title VI of the Civil Rights Act of 1964. Schools are required to respond in the short-term with discipline and in the long-term with a plan to manage the issue (U.S. Court of Appeals, 2012; McNeal, Kunkle, & Bryan, 2016). The Civil Rights Act of 1964 is a federal law to prevent bias toward race, color, religion, sex, or national origin (Public Law 88-352, 1964).

Some stakeholders argue that school district bullying policies restrict student speech of what can and cannot be posted on the Internet, as some accused of violating school policy claim their schools have violated their First Amendment on freedom of speech. "Overall, U.S. courts are oriented toward supporting First Amendment rights of free expression of students. Certain expressions, however, are not protected and allow intervention and discipline" (p. 3). These expressions include the disruption of the school learning environment by bullying/ harassment; impeding the educational process and school discipline; the obstruction of civil rights of other students; the use of school computers to harass others (Cyberbullying Research Center, 2015b). As state laws on bullying are adopted, courts are playing a role defining what schools are legally permitted to do to protect students, with some legal constraints presenting a barrier to state adoption of cyberbullying laws. Recent language in federal bills, such as the Safe Schools Improvement Act of 2017, have a specific provision reinforcing that the bill does not alter legal standards or federal laws protecting the freedom of speech or expression (House of Representatives 1957, 2017).

Judicial decisions have influenced subsequent court decisions related to cyberbullying. For example, in case law *Tinker v. Des Moines Independent Community School District*, a precedent court decision was used to regulate cyberbullying (U.S. Supreme Court, 1969). In Tinker, the Court ruled that the school violated the First Amendment rights of those students who wore black armbands to school as a symbolic protest to the Vietnam War. The Court ruled that a student's right to free speech was not absolute, but to restrict student speech or actions school officials must show that student actions violated the rights of other students or constitute a "substantial disruption" to the school's ability to maintain order. Courts have extended the substantial disruption standard of the Tinker decision to cyberbullying. Since cyberbullying differs from other types of bullying in that it can occur 24/ 7 and can reach a bigger audience than traditional offline bullying, federal courts have been extending the Tinker standard to include actions that have taken place off school grounds. For example, in *Doninger v. Niehoff,* a student posted on their public blog that a school event was cancelled and urged school peers to harass the school administrator. The Second Circuit Court agreed on the school's discipline to bar the student from participating in student government, deciding that the student's blog constituted a substantial disruption (U.S. Court of Appeals, 2008). In all, court cases have played a role in shaping cyberbullying policy for federal, state and school districts. All three branches of government are involved on bullying issues by passing and shaping legislation, and having a check and balance on each other.

Government Initiatives Protecting the LGBT Population

Crimes toward lesbian, gay, bisexual, and transgender (LGBT) persons occur every hour in the U.S. (Human Rights Campaign, 2017b). According to the 2015 Gay, Lesbian & Straight Education Network's National School Climate Survey, 20% of LGBT students reported bias due to gender expression (House Concurrent Resolution 49, 2017). Congress declares violence motivated by actual or perceived gender, gender identity, or sexual orientation a serious issue (Public Law 111-84, 2009). In response, federal laws have been passed. Landmark is the Civil Rights Act of 1964 (Public Law 88-352) signed into law by former President Lyndon Johnson. This Act aims to enforce constitutional rights to vote, "confer jurisdiction upon the district courts of the United States to provide injunctive relief against discrimination," to prevent discrimination in public accommodations and institutions and in government funded programs toward race, color, religion, sex, or national origin (Public Law 88-352, 1964). Although there is some contention among citizens on the meaning of the Act's use of the word "sex" as some claim it has meaning beyond one's biological sex, the 1964 Act is considered a foundation for subsequent laws and court cases.

Responsibility to prohibit aggression (hate) crimes toward the LGBT persons is largely being placed on institutions receiving federal monies. Those institutions, such as schools receiving government funding are required to comply with government anti-LGBTQ aggression initiatives. According to the government website Stopbullying.gov, "when students are harassed based on their actual or perceived sexual orientation, they may also be subjected to forms of sex discrimination recognized under Title IX" (2014). Title IX of the Education Amendments of 1972 is a federal administrative law on educational institutions receiving federal dollars to prohibit sex discrimination. It applies to primary, secondary, colleges, universities and educational training programs in the U.S. (USDOJ, August 2015). Newer laws follow the 1972 idea on anti-harassment and anti-discrimination of LGBT persons. For example, the federal Don't Block LGBTQ Act of 2016, was introduced in the House to prohibit primary and secondary education schools or libraries receiving "discount rates" for Internet from blocking Internet access to lesbian, gay, bisexual, transgender, or queer (H.R. 6254, 2015-2016). The 2017 Safe Schools Improvement Act, in committee, is a bipartisan bill requiring school districts receiving federal dollars to adopt codes against bullying and harassment and against bias of race, color, national origin, sex, sexual orientation, gender identity, disability, and religion (H.R. 1957, 2017-2018). In addition to schools who receive monies, are the recipients of the VAWA grants. The VAWA of 2013 with its LGBT provision is aimed at "discrimination for LGBT people appearing within a federal funding statute." The VAWA Culturally Specific Services Program gives LGBT victims protection and services (USDOJOVW, 2016, p. 54).

Laws protecting the LGBT population from bias are not limited to cyberspace, school settings, and those institutions receiving federal monies. According to the National Conference of State Legislators, many states have passed laws similar to the federal civil rights acts protecting vulnerable populations from bias in employment and public accommodation. California for example includes sex, gender identity, gender expression, sexual orientation in their law against both public and private employment discrimination. To this list Colorado adds victims of stalking, domestic violence and sexual assault (NCSL, 2017). More than 200 U.S. cities and counties have passed policy banning gender identity discrimination. Transgender persons as public employees may be protected through ordinances, charter provisions and like (American Civil Liberties Union, 2017).

CONCLUSION

All three branches of U.S. government to some extent have had a voice in today's cyber policies. Overall, federal policy fighting cyber aggression has been incremental with the "cyber" added later to older family domestic violence laws protecting women and children against physical aggression. The federal government, all 50 states, the District of Columbia, and U.S. territories have adopted some type of policy addressing stalking crime (USDOJOVW, 2017). However, fast paced digital advances used by criminals and a slow legislative process creates a "lag" in the making of modern cyber laws. Not all states have picked up where the federal government has left off in passing strict protective regulations and accompanying administrative policy. While some states are passing specific laws differentiating cyber aggression from its physical offline form, other states are laggard by either passing a few law(s) having broad language to cover the evolving methods of cyber aggression, or relying on existing cyber laws to cover multiple types of digital aggression. In all, state laws are not uniform and vary in extent of protective coverage and perpetrator penalty.

Cyberbullying tends to be an issue for younger aged Americans although older adults can also face this type of aggression. The federal governing authority over cyberbullying has largely been devolved to the states and school districts, but the feds still have a hand in the matter requiring those schools receiving federal school funding to take aggression seriously. Responsibility to prohibit aggression (hate) crimes toward the LGBT population is also largely being placed on institutions receiving federal monies. Several 2017 federal bills have been introduced to protect the LGBT person(s). The magnitude of state actions reflecting the evolving political and policy environment relative to bullying in schools suggests that legislators are repeatedly refining legislative expectations for schools in response to emerging problems such as cyber bullying (U.S. Dept. of Education, 2011). U.S. Courts have played a role in shaping state bullying policy. In the next chapter, community anti-bullying programs are discussed.

REFERENCES

American Civil Liberties Union. (2017). *Know your rights: transgender and the law*. Retrieved June 15, 2017, from https://www.aclu.org/know-your-rights/transgender-people-and-law

Antibullying Institute. (2017). *Bullying facts and the challenge to be met*. Retrieved June 23, 2017, from http://antibullyinginstitute.org/facts#.WS8Tdfn1DIU

Baumgartner, F. R., & Jones, B. D. (1993). *Agendas and instability in American politics*. Chicago: The University of Chicago Press.

Black, H. (1990). *Black's law dictionary* (6th ed.). St. Paul, MN: West Group.

Cobb, R., & Elder, C. (1971). The politics of agenda setting. *The Journal of Politics*, *33*(4), 892–915. doi:10.2307/2128415

Congressional Research Service. (2014). *The violence against women act: overview, legislation, and federal funding*. Retrieved June 23, 2017, from https://www.crs.gov

Congressional Research Service. (2015). *The violence against women act: overview, legislation, and federal funding*. Retrieved June 23, 2017, from https://www.crs.gov

Cyberbullying Research Center. (2015a). *State cyberbullying laws: A brief review of state cyberbullying laws and policies*. Retrieved June 5, 2017, from https:// from www.cyberbullying.org

Cyberbullying Research Center. (2015b). *Cyberbullying legislation and case law implications for school policy and practice*. Retrieved June 5, 2017, from https://www.cyberbullying.org

Cyberbullying Research Center. (2015c). *State sexting laws*. Retrieved June 5, 2017, from https://www.cyberbullying.org

Cyberbullying Research Center. (2016). *State cyberbullying laws. a brief review of state cyberbullying laws and policies*. Retrieved June 5, 2017, from https://www.cyberbullying.org

Cyberbullying Research Center. (2017). *Bullying laws across America*. Retrieved June 5, 2017, from https://www.laws.cyberbullying.org

Dye, T. R. (2008). Understanding public policy (12th ed.). Upper Saddle River, NJ: Pearson Prentice Hall.

Federal Bureau of Investigation Internet Crime Complaint Center. (2017). *2016 Internet Crime Report*. Retrieved December 29, 2017, from http://www.ic3.gov/media/annualreports.aspx

Goodno, N. H. (2007, Winter). Cyberstalking, a new crime: Evaluating the effectiveness of current state and federal Laws. *Missouri Law Review*, *72*(1), 125–197.

House Concurrent Resolution 49. (2017). *Supporting the goals and ideals of GLSEN's 2017 Day of Silence in bringing attention to anti-lesbian, gay, bisexual, transgender, and queer name-calling, bullying, and harassment faced by individuals in schools. 115th Congress, 1st Session*. Retrieved July 10, 2017, from https://www.govtrack.us/congress/bills/115/hconres49

House of Representatives 1957. (2017-2018). *Safe Schools Improvement Act of 2017*.

House of Representatives 2151. (June 2017). *Tyler Clementi Higher Education Anti-Harassment Act of 2017*.

House of Representatives 4123. (1995-1996). *Child Pornography Prevention Act of 1996. 104th Congress.*

House of Representatives 6254. (2015-2016). *Don't Block LGBTQ Act of 2016*.

Human Rights Campaign. (2017). *Federal legislation*. Retrieved June 23, 2017, from http://www.hrc.org/resources/federal-legislation

Human Rights Campaign. (2017b). *Hate crimes law 2017*. Retrieved June 23, 2017, from http://www.hrc.org/resources/hate-crimes-law

Legislature, F. (2015). *Senate Bill 538*. Retrieved December 26, 2017, from https://www.flsenate.gov/Session/Bill/2015/0538/BillText/er/PDF

Lonardo, T., Martland, T., & White, D. (2016). A legal examination of revenge pornography and cyber-harassment. *Journal of Digital Forensics, Security and Law.*, *11*(3), 79–105.

Lowi, T. (1972). Four systems of policy, politics and choice. *Public Administration Review*, *33*(4), 298–310. doi:10.2307/974990

McNeal, R., Kunkle, S., & Dotterweich-Bryan, L. (2016). State-Level Cyberbullying Policy: Variations in Containing a Digital Problem. In G. Crews (Ed.), *Critical Examinations of School Violence and Disturbance in K-12 Education* (pp. 62–82). Hershey, PA: Information Science Reference. IGI Global. doi:10.4018/978-1-4666-9935-9.ch005

McNeal, R., & Schmeida, M. (2015). Digital Paranoia: Unfriendly Social Media Climate Affecting Social Networking Activities. In J. P. Sahlin (Ed.), *Social Media and the Transformation of Interaction in Society* (pp. 210–227). Hershey, PA: Information Science Reference. IGI Global. doi:10.4018/978-1-4666-8556-7.ch011

McNeal, R., Tolbert, C., Mossberger, K., & Dotterweich, L. J. (2003). Innovating in digital government in the American states. *Social Science Quarterly*, *84*(1), 52–70. doi:10.1111/1540-6237.00140

National Conference of State Legislators. (2013b). *State bullying legislation since 2008*. Retrieved June 5, 2017, from http://www.ncsl.org/research/education/bullying-legislation-since-2008.aspx

National Conference of State Legislators. (2017). *State laws on employment-related discrimination*. Retrieved June 5, 2017, from http://www.ncsl.org/research/labor-and-employment/discrimination-employment.aspx

National Conference of State Legislatures. (2010). *Cyberbullying and the states*. Retrieved June 5, 2017, from http://www.ncsl.org/research/civil-and-criminal-justice/cyberbullying-and-the-states.aspx

National Conference of State Legislatures. (2011). *Overview: School bullying*. Retrieved June 5, 2017, from http://www.ncsl.org/research/education/school-bullying-overview.aspx

National Conference of State Legislatures. (2013a). *Cyberstalking and cyber harassment laws*. Retrieved June 5, 2017, from http://www.ncsl.org/research/telecommunications-and-information-technology/cyberstalking-and-cyberharassment-laws.aspx

National Institute of Justice. (1996). *Domestic violence, stalking, and antistalking legislation – an annual report to congress under the violence against women act* (NCJ 160943). Retrieved June 1, 2017, from https://www.ncjrs.gov/pdffiles/stlkbook.pdf

Pew Internet & American Life Project. (2013). *Anonymity, privacy, and security online*. Retrieved July 10, 2017, from http://pewinternet.org/Reports/2013/Anonymity-online.aspx

Pew Research Center. (2015). *Mobile messaging and social media 2015*. Retrieved July 10, 2017, from http://www.pewinternet.org/2015/08/19/mobile-messaging-and-social-media-2015/

Pew Research Center. (2017). *The Internet of things connectivity binge: what are the implications?* Retrieved July 10, 2017, from http://www.pewinternet.org/2017/06/06/the-internet-of-things-connectivity-binge-what-are-the-implications/

Privacy Rights Clearinghouse. (2016). *Online harassment & cyberstalking*. Retrieved June 5, 2017, from https://www.privacyrights.org

Public Law 103-322. (1994). *Violent Crime Control and Law Enforcement Act of 1994.*

Public Law 105-314. (1998). Protection of Children from Sexual Predators Act of 1998.

Public Law 106-386. (2000). *Trafficking and Violence Protection Act of 2000.*

Public Law 109-162. (2005). *Violence against Women and Department of Justice Reauthorization Act.*

Public Law 111-5. (2009). *American Recovery and Reinvestment Act of 2009.*

Public Law 111-84. (2009). *Matthew Shepard and James Byrd, Jr. Hate Crimes Prevention Act.*

Public Law 113-4. (2013). *Violence against Women Reauthorization Act of 2013.*

Public Law 88-352. (1964). *Civil Rights Act of 1964.*

Public Law 89-10. (1965). *Elementary and Secondary Education Act of 1965.*

Public Law 90-100. (1967). *Act Creating the Commission on Obscenity and Pornography.*

Public Law 90-351. (1968). *Omnibus Crime Control and Safe Streets Act of 1968.*

Public Law 95-225. (1978). *Protection of Children against Sexual Exploitation Act.*

Public Law 98-292. (1984). *Child Protection Act of 1984.*

Public Law 98-457. (1984). *Family Violence Prevention and Services Act.*

Ripley, R. B., & Franklin, G. A. (1980). *Congress, the bureaucracy, and public policy*. Homewood, IL: The Dorsey Press.

Stalking Resource Center. (2015). *Stalking fact sheet.* Retrieved June 23, 2017, from https:// www.victimsofcrime.org/src

Stalking Resource Center. (2017). *Stalking resource center services*. Retrieved June 23, 2017, from https://victimsofcrime.org/our-programs/stalking-resource-center/about-us

State of Illinois Public Act 098-1138. (2015). *An Act Concerning Criminal Law.*

Stopbullying.gov. (2014). *Federal laws*. Retrieved June 15, 2017, from https:// www.stopbullying.gov/laws/federal/index.html

U. S. Department of Health & Human Services. (2015). *What is cyberbullying?* Retrieved January 5, 2015, from http://www.stopbullying.gov/

U. S. Department of Justice Office on Violence Against Women. (2016a). *The 2016 biennial report to congress on the effectiveness of the grant programs under the violence against women act.* Retrieved June 15, 2017, from http:// www.ovw.usdoj.gov

University of North Carolina at Chapel Hill. (2017a). *Cyberstalking federal criminal statutes*. Retrieved June 5, 2017, from http://cyberstalking.web.unc.edu/federal-criminal-statutes/

University of North Carolina at Chapel Hill. (2017b). *Cyberstalking policy reform*. Retrieved June 5, 2017, from http://cyberstalking.web.unc.edu/policy-reform/

Urban Institute Justice Policy Center. (2013). *Citywide model bullying prevention policy*. Retrieved July 5, 2017, from https://www.urban.org

U.S. Court of Appeals, Second Circuit. (2008). *Doninger v. Niehoff*. 527 F.3d 41.

U.S. Court of Appeals, Second Circuit. (2012). *Zeno v. Pine Plains Central School District*. 702 F.3d 655.

U.S. Department of Education, Office of Planning, Evaluation and Policy Development, Policy and Program Studies Service. (2011). *Analysis of state bullying laws and policies*. Retrieved June 23, 2017, from http://www.ed.gov/about/offices/list/opepd/ppss/index.html

U.S. Department of Justice. (2015). *Overview of title IX of the education amendments of 1972, 20 U.S.C. A§ 1681 ET. SEQ*. Retrieved June 5, 2017, from https://www.justice.gov/crt/overview-title-ix-education-amendments-1972-20-usc-1681-et-seq

U.S. Department of Justice Office of Justice Programs. (2012). *Stalking victims in the United States – revised - Bureau of Justice statistics special report* (NCJ 224527). Retrieved June 23, 2017, from http://www.bjs.gov/content/pub/pdf/svus_rev.pdf

U.S. Department of Justice Office on Violence Against Women. (2015a). *2014 tribal consultation report*. Retrieved June 5, 2017, from http://www.ovw.usdoj.gov

U.S. Department of Justice Office on Violence Against Women. (2015b). *STOP program 2014 report part A*. Retrieved June, 15, 2017, from http://www.ovw.usdoj.gov

U.S. Department of Justice Office on Violence Against Women. (2016b). *Report to congress on the 2013-2015 activities of grantees receiving federal funds under the grants to reduce violent crimes against women on campus program*. Retrieved June 15, 2017, from http://www.ovw.usdoj.gov

U.S. Department of Justice Office on Violence Against Women. (2017). *2014 report to congress grant funds used to address stalking.* Retrieved June 15, 2017, from http://www.ovw.usdoj.gov

U.S. Supreme Court. (1969). *Tinker v. Des Moines Independent Community School District.* 393 U.S. 503.

47. USC § 223. (2012). *Obscene or Harassing Telephone Calls in the District of Columbia or in Interstate or Foreign Communications.*

18. USCS § 2261A. (2012). *Stalking.*

18. USCS § 875. (2012). *Interstate Communications.*

Chapter 4
Community Programs:
Local School Boards and Anti-Bullying Programs

ABSTRACT

The United States has a federal system. One advantage of a federal system is that it can encourage competition among the states resulting in the testing of new policy solutions and the diffusion of best practices. This holds true for online aggression policy, particularly those addressing cyberbullying. This chapter begins with a discussion of the literature on strategies being adopted at the school board level to limit the spread of cyberbullying. It concludes with an overview of current evaluation research comparing recent policies being implemented by local schools.

INTRODUCTION

As illustrated by Chapter 3, there has been considerable legislative efforts at both the federal and state levels to combat online aggression. While policy adoption is crucial to solving any public problem, the policy process does not stop there. Programs once adopted must be implemented and those adopted at the state-level to curb cyber aggression have often been left to local law enforcement and school districts to carry out. It is not that unusual for local governmental agencies to be given considerable leeway in how they will implement policy. A typical state-level anti-bullying policy may direct school

DOI: 10.4018/978-1-5225-5285-7.ch004

districts to have a policy in place to address both traditional and cyberbullying but may not provide any guidance on which preventive measures to use. This has resulted in significate differences in anti-bullying programs, not only across the nation, but intrastate. The effectiveness of each program also varies greatly. One positive result of this variation in programs is that the school districts are acting as "laboratories of democracy." Because many different programs have been implemented, we are able to compare them to help develop a list of best practices. This chapter looks at the contribution of local government in the battle against cyber aggression through the role of local school districts in implementing anti-bullying policy. It begins with an overview of the impact of bullying, followed by a discussion of anti-bullying programs in the United States. It will conclude with a presentation of research into the effectiveness of different intervention strategies.

BULLYING AS A PUBLIC HEALTH PROBLEM

The research speaks very clearly and with little ambiguity, in stating that the impact of bullying and its cyber companion is linked to many negative outcomes such as mental health issues, substance abuse, and suicide (Institute of Medicine [IOM] & National Research Council [NRC], 2014; Ttofi & Farrington, 2008, 2012; Polanin, Espelage, & Pigott, 2012; Olweus, 1993, Olweus, Limber & Mihalic, 1999). Bullying is associated with anxiety, depression, failing or near failing school performance, and delinquent behavior. The act of bullying can manifest itself in derogatory comments and name calling; social exclusion and isolation; hitting, kicking, shoving and spitting; lying and spreading false rumors; having personal items and money taken and/or damaged; being threatened or forced to do something not wanted; and, negative actions directed at a another because of race, ethnicity, national origin, religion, gender expression, gender identity, sexual orientation, or disability (Hazelden, 2016; IOM & NRC, 2014; Olweus, 2013).

Victimization can and does occur in all age groups ranging from children younger than elementary school age and adolescents and young adults high school age and beyond (IOM & NRC, 2014). Students who attend schools with high incidences of bullying have lower grades than students at schools with less bullying (Strom, Thoresen, Wentzel-Larsen, & Dyb, 2013). Poor academic outcomes can be attributed to anxiety, inability to concentrate, and attendance problems (Lee & Cornell, 2009). Participation in bullying, whether as a perpetrator or victim, has long term consequences on the physiological and psychological heath of children and their psychosocial adjustment as adults (Ttofi & Farrington, 2008). For example, headaches, stomach aches, dizziness, bedwetting, and sleep disturbance all may be warning signs that bullying is occurring (Substance Abuse & Mental Health Services Administration [SAMHSA], 2014).

The precise rate of bullying victimization and offending varies according to age, gender, and type of bullying as well as the length of time over which bullying behaviors are assessed and by subgroup (Hertz, Donato, & Wright, 2013). According to Indicators of School Crime and Safety (2016), younger children are more likely to be victimized by bullying than older adolescents. Females (23%) tend to be bullied more than males (19%). Additionally, the type of bullying experienced by females and males differs. Females tend to be made fun of, called names, become the subject of rumors, and are socially isolated and excluded from activities. Males tend to be pushed, shoved, kicked, spit on, and threatened with harm (Musu-Gillette, et. al., 2016). Specific subgroups are more likely to be victimized. For example, youth who identify as a sexual minority are more likely to be victimized than their heterosexual companions. In 2015, 34% of gay, lesbian, or bisexual students in grades 9 through 12 reported that they had been bullied during the previous 12 months as compared with 19% of heterosexual youth (Musu-Gillette, et. al., 2016). LGBTQ students who experience victimization and discrimination at school have inferior educational outcomes and poorer psychological health (Kosciw, et al., 2016). Data from the 2015 National Youth Risk Behavior Survey (YRBS) indicate that more than 1 in 10 lesbian, gay, and bisexual students reported missing school during a 30 day span due to safety concerns (Centers for Disease Control and Prevention [CDC], 2016). Additionally, more than 40% of lesbian, gay, bisexual students have seriously considered suicide and 29% reported having attempted suicide during the past 12 months (CDC, 2016).

According to the Olweus Bullying Prevention Program, the impact of bullying can have significant consequences on the student victim, the student who bullies, students that observe the bullying, and the overall climate of the school and community (Hazelden Foundation, 2016). Students who are bullied can exhibit characteristics of depression, low self-esteem, health problems, academic failure, truancy, substance use and suicidal ideation. Students who bully others are more likely to steal and vandalize property, consume alcohol, smoke cigarettes, fight, carry a weapon, and fail in school. Students who observe others being bullied tend to be fearful, feel impotent to act and guilty for not acting, and, at times, are tempted to participate in the bullying. Finally, schools with bullying problems develop a culture of fear and disrespect for others. In such a climate, students feel insecure, dislike school, and have difficulties concentrating in class (Hazelden Foundation, 2016).

Victims of bullying have been found to have elevated internalizing behaviors such as depression, anxiety, withdrawal, and avoidance; engage in externalizing behaviors; possess negative self-related cognitions; and lack adequate social skills. The youth who is victimized is rejected by peers and socially isolated. Bully-victims have both internalizing and externalizing behaviors and negative attitudes and beliefs about themselves. The youth who is a bully-victim has inadequate problem-solving skills, is not successful in school, rejected by peers and socially isolated, and negatively influenced by peers (Cook, et al., 2010; Espelage & Holt, 2013). As stated in a report issued by the National Academies of Sciences, Engineering, Medicine (2016), bullying has both short, as well as, long term internalizing and externalizing consequences for those who are involved in bullying behaviors. Internalizing factors can manifest in psychosomatic complaints, suicidal ideation, and mental health issues. Externalizing factors can manifest in heightened aggression, social isolation and rejection, conduct problems, and withdrawal from school. Additionally, the Committee concluded that although the effects of being bullied on the brain are not yet fully understood, there are changes in the stress response system in the brain, that are associated with an increased risk for mental health problems, cognitive function, self-regulation, and other physical health problems (2016).

SUICIDE AND BULLYING

Suicide, as defined by the Centers for Disease Control and Prevention, is self-directed violence, including acts of suicidal behavior, both fatal and nonfatal attempts; suicidal ideation or thinking about, considering, or planning for suicide; and non-suicidal intentional self-harm or behaviors with the intention not to kill oneself, as in self-mutilation (CDC, n.d.). In 2013, suicide was the second leading cause of death for the age group 10–24 and accounted for 16.8% of deaths in that age group (Heron, 2016). Suicide related behavior is complicated and rarely the result of one trauma. Youth who are at risk usually are dealing with a complex interaction of multiple relationships, mental health issues, and school and home stressors (CDC, 2016). According to the American Foundation for Suicide Prevention (AFSP), suicide is the result of multiple factors and most frequently occurs when stressors exceed the coping abilities of someone suffering from a mental health condition

(2017). Disorders like depression, anxiety, substance abuse problems, and prolonged stress, especially when neglected, increase the risk for suicide (AFSP, 2017). Prolonged stress factors may include harassment, bullying, relationship problems, and unemployment (AFSP, 2017). Depression is the most common condition associated with suicide (CDC, 2016).

As has been indicated throughout this chapter, bullying is a significant public health problem that has short- and long-term physical and mental consequences for children and adolescents (National Academies of Science, Engineering and Medicine, 2016; CDC, 2016; Hertz, Donatoo, Wright, 2013). Involvement in bullying, in any capacity, is related to an increased risk for suicide and suicide-related behaviors (CDC, 2016; Espelage & Holt, 2013). Thus, youth who are involved in bullying are more likely to report higher levels of suicide related behavior in comparison with youth who are not impacted by bullying (CDC, 2016). Other studies find a similar relationship bullying and suicide. For example, Borowsky, Taliaferro, and McMorris (2013) analyzed data on 130,908 students who responded to the Minnesota Student Survey and were enrolled in the sixth, ninth, and twelfth grades. The study identified students who were involved in frequent bullying (once a week or more during the past 30 days) and compared responses between those who did and did not report suicidal ideation or a suicide attempt during the past year. Separate analyses were conducted for perpetrators only, victims only, and bully victims. Slightly over six percent (6.1%) of the respondents reported frequent bullying only, 9.6% frequent victimization only, and 3.1% both. Suicidal ideation and/or a suicide attempt were reported by 22% of perpetrators only, 29% of victims only, and 38% of bully-victims. Risk factors that were pertinent to all three bullying groups were a history of self-injury and emotional distress. Physical abuse, sexual abuse, a mental health problem, and running away from home were additional risk factors for perpetrators only and victims only. Parental connectedness was a protective factor for all three bullying groups (Borowsky, Taliaferro, & McMorris, 2013).

SCHOOL SHOOTINGS

School shootings are a rare, but significant, component of school violence in America. Between 1996 and 2008, 22 incidents occurred in which students started firing at schoolmates and teachers at their school or at a school-sponsored event. These incidents, including the Columbine High School incident, left 85 dead and 112 injured (Newman, et al., 2008). Although sometimes lost in the discussions on bullying, it was the school shooting at Columbine that resulted in the passage of the first anti-bullying programs in the United States. Following the attack at Columbine High School, the U.S. Secret Service and the U.S. Department of Education launched a collaborative effort to study school shootings. The project was called the Safe School Initiative. The Safe School Initiative was an extensive examination of 37 incidents of targeted school shootings involving 41 attackers that occurred in the United States beginning with the earliest identified incident in 1974 through May 2000. The focus of the Initiative was on examining the thinking, planning, and other actions engaged in by students who carried out school attacks (Vossekuil, et al., 2004). The study concluded that there was no accurate psychological or behavioral profile of the school attacker that, in fact, the personality and social characteristics of the shooters varied significantly. However, the study also found that in several cases bullying played a key role in the decision to attack (2004).

According to the report, slightly less than three-quarters (29 or 71%) of the attackers felt that they had been bullied, persecuted, or injured by others prior to the attack. In several cases, individual attackers had experienced bullying and harassment that was long-standing and severe. In some of these cases the experience of being bullied seemed to have a significant impact on the attacker and appeared to have been a factor in his decision to attack the school. In one case, many of the attacker's schoolmates described the attacker as "the kid everyone teased" (Vossekuil, et.al. 2004). Witnesses (schoolmates) alleged that nearly every child in the school had at some point thrown the attacker against a locker, tripped him in the hall, held his head under water in the pool, or thrown things at him.

The National Academies of Sciences, Engineering, Medicine (2016) in a review of the evidence found that most of the studies into school shootings have concluded that prior bullying, with an emphasis on the shooter being the target of bullying, may have played a role in many school shootings but not all. While bullying was identified as a factor, it did not appear to be the most salient factor in the decision to engage in these violent crimes. Additionally, there was not enough qualitative and/or quantitative evidence to conclude that bullying was a causal factor in multiple-homicide targeted school shootings nor was the evidence precise with regards to the interplay between bullying and mental health issues. The Committee concluded the data are unclear on the role of bullying as one of, or a precipitating cause of school shootings (The National Academies of Sciences, Engineering, Medicine, 2016).

THE GOAL IS INTERVENTION

How do we address the problem of bullying (both its traditional and cyber forms)? According to the Centers for Disease Control and Prevention, the goal in managing the spectrum of bullying behaviors is prevention (2016). In other words, to stop bullying before it begins (CDC, 2016). Labelled a public health problem, the Centers for Disease Control and Prevention, propose a process consisting of four steps. The first step is to define and monitor the behavior which includes understanding the scope of the problem and who it affects through research and study. The second step is to identify the risk and protective factors and develop programs that negate the risk factors and validate the protective factors. The third step is to develop and test prevention strategies predicated upon the evidence gained in research and study. The last step is to facilitate the adoption by communities, schools, youth groups, and others of evidence based prevention and intervention strategies that address bullying behaviors (2016). Ideally, all anti-bullying school programs would prevent bullying before it starts. That may not be realistic, but school programs can at least minimize the prevalence of bullying. As discussed by Perren et al. (2012), anti-bullying programs should have three components: elements to minimize the risks of bully; those that combat bullying when it occurs and those that minimize its effects once it has occurred (p. 284).

CATEGORIES OF ANTI-BULLYING INTERVENTION PROGRAMS

Most anti-bullying programs are school-based and according to Limber (2003) predicated on an awareness that most bullying does not involve physical violence. Bullying exists in a relationship in which there is a power imbalance, and occurs repeatedly over time. Typically, intervention programs fall into six categories: educational or awareness programs; zero tolerance programs; therapeutic programs; mediation and conflict resolution; curricular programs; and comprehensive programs designed to reflect the unique needs of a school and community. Not all intervention programs are equally effective. Even though school officials may instinctively build educational or awareness components into their anti-bullying programs, there is little evidence that they are effective. While research by Limber (2013) found that awareness raising efforts did educate and inform, the research did not find awareness raising efforts to be effective at changing cultural norms and bullying behavior. Furthermore, other studies failed to provide support for the argument that awareness raising events or assemblies were effective at changing a climate of bullying or producing long-term positive effects on bullying behavior (National Academies of Sciences, Engineering, and Medicine, 2016; Farrington & Ttofi, 2009).

Anti-bullying programs that rely on education or awareness programs are not the only ones to have mixed reviews. School expulsion and exclusion interventions are defined as zero tolerance programs that mandate removing the bully from the school environment. Zero tolerance policies and practices frequently are the default response by school staff and administrators in bullying situations. However, research suggests that these approaches do not appear to be effective and may result in increased academic and behavioral problems for youth (National Academies of Sciences, Engineering, and Medicine, 2016; American Psychological Association Zero Tolerance Task Force 2008; Skiba, et al., 2006). Mediation and conflict resolution are positive approaches for dealing with conflict between two or more individuals. However, bullying is more about victimization and less about conflict. Programs using mediation and/or conflict resolution techniques are premised on the assumption that both parties involved in the conflict share blame. In other words, according to the philosophy behind mediation and conflict therapy, the victim is just as responsible for his or her victimization as is the bully. Conflict resolution approaches suggest a disagreement between two peers of equal status or power, rather than an instance of peer abuse. Actions that tend to blame the victim can contribute to further victimization of the victim (Limber, 2003). Also, some peer-mediated conflict resolution programs seem to increase attitudes supportive of bullying, whereas others showed an increase in incidents of targeting rather than a reduction in bullying-related behaviors (National Academies of Sciences, Engineering, and Medicine, 2016; Farrington & Ttofi, 2009).

Other strategies for addressing bullying have found to be more effective. Therapeutic treatment programs for youth identified as bullies may include anger management classes, programs focusing on raising self-esteem, social-emotional learning, or skill building. Some mental health counseling programs, such as cognitive behavioral therapy, have been demonstrated as successful with certain populations (National Academies of Sciences, Engineering, and Medicine, 2016) and some are important components of a comprehensive school based approach to bullying (Farrington & Ttofi, 2009; Limber, 2003). In a meta-analysis of school-based mental health promotion programs Durlak, et al (2011), found such programs to improve social-emotional skills, pro-social norms, school bonding, positive social behavior, and to result in reduced problem behaviors, such as aggression, substance use, and internalizing symptoms (Durlak et al., 2011).

In addition, numerous curricular approaches and curricula have been developed for use in schools and have been found to reduce incidences of bullying. In general, these programs try to explain bullying and its effects, teach strategies to avoid bullying or for intervening in bullying incidents, and build social cohesion among students. Some of these programs have been evaluated, and some evidence-based are found to be effective in improving desired outcomes (Finkelhor, et al., 2014; Durlak, et al., 2011; Vreeman & Carroll, 2007). According to research conducted by Seeley, Tombari, Bennett, and Dunkel (2011), it is central that curricular approaches are designed to address the needs of a school or community; simply dropping mass-produced programs into schools rarely works. Comprehensive approaches target the larger school community to change school climate and norms. They acknowledge the need for a long-term commitment to addressing bullying, but often do so as part of a larger violence prevention effort (Limber, 2003). Comprehensive programs may incorporate the expertise and involvement of teachers, education support professionals, school resource officers, families, health care professionals, and community members, thereby attempting to support youth across multiple ecological levels (National Academies of Sciences, Engineering, and Medicine, 2016). Systematic reviews and meta-analyses have consistently found that the prevention programs that are considered effective tend to be a whole school multi-component program(s) combining elements of universal and targeted strategies (National Academies of Sciences, Engineering, and Medicine, 2016; Durlak & Weissberg, 2013; Farrington & Ttofi, 2009; Ttofi, Farrington, & Baldry, 2008).

The Olweus Anti-Bullying Program is one of the most recognized example of a comprehensive approach to preventing bullying. It is the most extensively studied bullying prevention program in the world (National Academies of Sciences, Engineering, and Medicine, 2016). Farrington and Ttofi in a meta-analysis, found that programs conceptually based on the Olweus Bullying Prevention Program were the most effective, compared to the other programs they reviewed (2009). Developed by Dan Olweus of the University of Bergen in Norway, the Olweus Bullying Program is a prevention program involving individual, classroom, school wide, and community strategies that create a safe and positive school climate, improve peer relations, and increase awareness of and reduce the opportunities and rewards for bullying behavior (Blueprints for Healthy Youth Development, n.d.).

AN EVALUATION OF ANTI-BULLYING PROGRAMS

Davis and Nixon (2010)

The Youth Voice Project surveyed more than 13,000 students in 31 schools in the United States (2009-2010) to determine which actions were most helpful for bullied and mistreated youth. The authors developed an on-line survey instrument which was designed to assess school connectedness, varying forms of peer victimization such as physical and relational aggression, and strategies used by students to address the victimization (Davis & Nixon, 2010). Examples of physical aggression include hitting, shoving, spitting, and threatening. Examples of relational aggression include name calling, spreading rumors, isolation, and exclusion.

Youth were most likely to indicate that actions in which they reached out and received support from others, such as telling adults at school/home and/or friends, made a positive difference. Listening, providing advice and guidance, and follow up or checking back with the youth to determine if the situation had improved were identified by youth as specific examples of supportive actions by adults that made a positive difference. Actions taken by adults which resulted in a worsening of the situation were those which directed blame and recrimination at the student victim and included such things as admonishment for tattling, advising the student-victim to solve the problem her/himself, ignoring the bullying, and advising the student-victim that his/her behavior was at fault (Davis & Nixon, 2010).

Strategies that found to be least effective were those directed at the perpetrator's behavior including asking the offender to stop or informing the student-bully about the hurtfulness of his/her behavior. Strategies which intensified the situation, i.e. made it worse, included fighting back, striking back, making plans to retaliate, and doing nothing. In other word, actions directed at changing the behavior of the perpetrator were perceived as resulting in a worsening of the situation (Davis & Nixon, 2010). According to the results from the Davis and Shaw research, a significant number of students are being victimized by peer mistreatment. Groups such as those delineated by race, religion, sexual orientation, and disability are more severely treated and affected than other youth. Emotional support and connection are two strategies which help to facilitate positive outcomes. Actions which blame the student-victim, retaliate against the student-bully, or ignore the problem were not effective strategies and resulted in a worsening of the situation. Educational personnel and peers demonstrated positive support for the student-victim through connection, encouragement, affiliation, and listening. The impact of peer support was more powerful than that of staff support (Davis & Nixon, 2010).

Ttofi, Farrington, and Baldry (2008)

Ttofi, Farrington, and Baldry (2008) conducted a meta-analysis that assessed 59 school-based anti-bullying programs in various countries, including the United States. The study classified as one the most comprehensive worldwide studies on anti-bullying programs, compared experimental and control groups and relied on data accessed from questionnaires completed by students. Results from the meta-analysis indicated that, overall, school-based anti-bullying programs are effective in reducing bullying and victimization. Results indicated that bullying and victimization were reduced by about 17–23% in experimental schools compared with control schools. The most important program elements that were associated with a decrease in bullying were parent training, improved playground supervision, disciplinary methods, school conferences, information for parents, classroom rules, classroom management, and videos. In addition, the total number of elements, and the duration and intensity of the program for children and teachers, were significantly associated with a decrease in bullying. (Ttofi, Farrington, & Baldry, 2008).

Farrington and Ttofi (2010)

The Campbell Collaboration Crime and Justice Group conducted a systematic review and meta-analysis of the effectiveness of programs designed to reduce school bullying perpetration and victimization. In this study, Farrington and Ttofi went beyond their previous study (2008) by expanding the search for studies and by focusing only on programs that were designed to reduce bullying and aggressive behavior (2010). Forty-four, out of 622 reports, were selected for inclusion in the study. Findings showed that school-based anti-bullying programs are often effective, and that particular program elements were associated with a decrease in bullying and victimization. One program element (work with peers) was significantly associated with a decrease in victimization. The authors determined that, on average, bullying decreased by 20% - 23% and victimization by 17% – 20%. The most important program elements that were associated with a decrease in both bullying and victimization were parent training/meetings, disciplinary methods, the duration of the program for children and teachers and the intensity of the program for children and teachers. Regarding the design features, the programs worked better with older children and in Norway specifically. Older programs and those in which the outcome measure was two times per month or more also yielded better results (Farrington & Ttofi, 2009; 2010).

Durlak, Weissberg, Dymnicki, Taylor, and Schellinger (2011)

Durlak, Weissberg, Dymnicki, Taylor, and Schellinger (2011) conducted a meta-analysis of 213 school-based, universal social and emotional learning (SEL) programs involving 270,034 kindergarteners through high school students. The current meta-analysis differs in emphasis from previous research syntheses by focusing exclusively on universal school-based social-emotional development programs and evaluating their impact on positive social behavior, problem behaviors, and academic performance. Durlak, et al., defined social and emotional learning programs as those that included specific "instruction in processing, integrating and selectively applying social and emotional skills ... in appropriate ways" (Durlak et al., 2011, p. 3), as well as programs where adults model these skills and children have opportunities to practice using them in diverse situations such that "safe, caring learning environments" are established organization-wide (Durlak et al., 2011, p. 3). Outcome of the meta-analysis indicated that SEL programs yielded significant positive effects on targeted social-emotional competencies and attitudes about self, others, and school while also enhancing students' behavioral adjustment in the form of increased prosocial behaviors and reduced conduct and internalizing problems, and improved academic performance on achievement tests and grades. The authors suggested that policy makers, educators, and the public can contribute to the healthy development of children by supporting the incorporation of evidence-based SEL programming into standard educational practice (Durlak, et al., 2011, p. 1).

Seeley, Tombari, Bennett, and Dunkel (2011)

Seeley, Tombari, Bennett, and Dunkel (2011), researchers from the National Center for School Engagement, in conjunction with the Office of Juvenile Justice and Delinquency Prevention, initiated a quantitative study to examine the impact of bullying on student engagement, attendance, and achievement and two qualitative studies to explore instructional, interpersonal, and structural factors at school that affect the connection between bullying and school attendance. Based on the result of the three studies and an extensive review of the literature the authors recommended that anti-bullying programs be designed to increase student engagement through service learning opportunities, model caring behavior for students, offer mentoring programs, start prevention programs early, address the challenging transition between elementary and middle school, and refrain from using mass-produced curriculums that are not aligned to local conditions (Seeley, et. al., 2011). One form of mentoring program that has shown to have promise is based on intervention. Hawkins, Pepler, and Craig (2001) found that just as the presence of bystanders helps to escalate an incidence of bullying, intervention by bystanders can also quickly deescalate the situation. Mentoring in Violence Program (MVP) is an anti-bully program that has been in existence for the last 20 years and has operated in states including Massachusetts, Iowa and Colorado. The program trains student leaders and school personnel how to promote a positive school environment and how (as a bystander) to step in and defuse a bullying situation (MVP, 2017).

Durlak and Weissberg (2013)

Durlak and Weissberg (2013) reviewed 68 studies involving youth attending an afterschool program that had the specific goal of fostering personal and social development and compared those cases to nonparticipating control youth. Programs which focused exclusively on academic achievement were not included in the review. The reviewed programs were located across the country, operated in urban and rural areas and served school-aged youth between 5 and 18 years old. Durlak and Weissberg hypothesized that effective programs would use evidence-based practices for enhancing young people's personal and social skills. Four evidence based practices were identified as significant in successful outcomes. The first, dealt with fidelity to the model or if the program staff used a sequenced step-by-step training approach (S);

the second, emphasized active forms of learning by having youth practice new skills (A); the third, focused on the time and attention dedicated to skill development (F) and, the fourth was if program staff were clear in defining the skills they were attempting to promote (E). SAFE programs, or programs which incorporated all four evidence based practices, were associated with significant improvements in self-perceptions, school bonding and positive social behaviors; significant reductions in conduct problems and drug use; and significant increases in achievement test scores, grades and school attendance. Results indicated that afterschool programs that follow evidence-based skill training practices are beneficial interventions for youth. The authors concluded by stating that the practical implications of the results suggest that policy and funding should be focused on assisting more afterschool programs to develop evidence-based practices (Durlak & Weissberg, 2013).

Finkelhor, Vanderminden, Turner, Shattuck, and Hamby (2014)

The National Survey of Children's Exposure to Violence II (NatSCEV II) is a "non-experimental" study, designed to obtain up-to-date incidence and prevalence estimates of a wide range of childhood exposure to violence and related risk factors (Finklehor, et. al., 2014). The study involved a national sample of 4,503 children and youth ages one month to 17 years in 2011. Study In addition, phone interviews were conducted with a subset of 3,391children ages 5–17. Sixty-five percent of the school age children (5–17) had participated in a violence prevention program, 55% in the past year. Bullying was the most frequent of the five most common topics of such programs, with 55% of children and youth having participated in a bullying prevention program. Twenty-one percent had been exposed to sexual assault prevention programming and about a third had been exposed to dating violence prevention.

Based on information from children's most recent program exposure, a majority (59%) of the programs involved single day, not multi-day curricula. Most (72%) gave youngsters information to take home and 64% of children discussed the program with parents. However, only 40% gave children the opportunity to practice skills, and only 18% invited parents to come in for a meeting about the program. Large percentages of the programs covered healthy and respectful relationships, warning signs for unsafe situations, and conflict resolution skills. The most widespread content was the advice to disclose the victimization to an adult (88%). Most respondents (71%) rated the programs as very or somewhat helpful. Younger children (5–9) who had been exposed to higher quality prevention programs had lower levels of victimization. Yet, the association did not apply to older youth or youth exposed to lower quality programs. Disclosure to authorities was more common for children who participated in higher quality programs. According to the authors, the results of this study are consistent with likely benefits from violence prevention education programs. However, the results also suggest that only a limited number of prevention programs include efficacious components (Finkelhor, et al., 2014).

AN EVALUATION OF ANTI-BULLYING PROGRAMS: SUMMARY COMMENTS

Table 1 summarizes the evaluation of the studies presented in this chapter. Several findings stand out. The first is that there is not a one-size-fits-all strategy for addressing bullying (either traditional or online). Demographical, social, attitudinal and jurisdictional factors influence the likelihood of success of any given strategy. The second result that was striking was that the individual studies emphasize difference stakeholders including peers, teachers, and parents. This suggests that to address the problem of bullying, the larger school community must be part of the solution and not simply school officials.

CONCLUSION

This chapter highlights the importance of local government in addressing different forms of cyber aggression. Much of the cyber aggression policy passed at the federal and state level is implemented at the local level either through local law enforcement or by school districts. It is not unusual for local governmental agencies to be given discretion in how they carry out these policies. This has resulted in considerable variation in how local governmental agencies address problems such as cyberbullying. This means that citizens in different jurisdictions are not receiving the same level of protection. On the other hand, it has also allowed policymakers the opportunity to observe which programs work best and for best practices to emerge. A review of current school programs finds that there are three general comments. The

Table 1. Summary of evaluation studies

Source	Summary
Davis & Nixon (2010)	Emotional support and connection for the victim were found to be strategies which help to facilitate positive outcomes while actions taken to change the behavior of the bully worsen the situation.
Ttofi, Farrington, & Baldry (2008)	Actions most associated with a decrease in bullying were parent training, improved playground supervision, disciplinary methods, school conferences, information for parents, classroom rules, classroom management, and videos.
Farrington & Ttofi (2010)	Actions most associated with a decrease in both bullying and victimization were parent training/meetings and disciplinary methods.
Durlak, Weissberg, Dymnicki, Taylor, & Schellinger (2011)	Social and emotional learning (SEL) programs which focus on teaching processing, integrating and selectively applying social and emotional skills are helpful in reducing bullying.
Seeley, Tombari, Bennett, & Dunkel (2011)	Recommends that anti-bullying programs be designed to increase student engagement through service learning opportunities, model caring behavior for students, offer mentoring programs, etc.
Durlak & Weissberg (2013)	Recommends anti-bullying strategies that include afterschool programs that follow evidence-based skill training practices.
Finkelhor, Vanderminden, Turner, Shattuck, and Hamby (2014)	Argues that anti-bullying strategies include violence prevention education programs.

first are strategies meant to prevent bullying (both online and traditional) and include strategies that focus on a positive school climate and those that teach students violence prevention training. To prevent cyberbullying this training might include how to use the Internet safely. The second component consists of procedures that deal with traditional bullying and cyberbullying after it occurs. These actions can include school expulsion and exclusion interventions. They also include therapeutic treatment programs such as anger management classes, programs which focus on raising self-esteem, social-emotional learning, or skill building. The final component is strategies that help the victim of bullying/cyberbullying cope. These may include peer support programs and teaching coping strategies (Perren, et al., 2012).

The review of current anti-bullying programs suggests a number of conclusions. The first is that there is no 'best' program that can be simply installed into a school system. Although the Olweus Anti-Bullying Prevention Program has been very successful in Norway, it not always worked as well in the United States. Research (Smith, et al., 2004) argues that the Olweus program works well in Norway because it was implemented following a series of suicides that were heavily covered by the media. Demographical, attitudinal and social differences of a jurisdiction must be factored in when developing a program. Another finding is that whatever program is adopted, it must have the commitment of stakeholders to see that implementation is followed through. In a number of high profile cases of teenage suicide, a commonality was that teachers, administrators and other school personal knew about instances of ongoing bullying but failed to do anything about it. A final conclusion is that a program is more likely to when other stakeholders besides school personal are included in process; included in this group of stakeholders should be parents. Research (Agatston, Kowalski, & Limber, 2007) finds that students are more likely to report being bullied to their parents instead of school personnel in part because they don't feel adults at school can help them. In addition, research (Berson, Berson, & Ferron, 2002) finds that youth are less likely to take part in risky online activities if they have a positive relationship with their parents that includes ongoing discussions about Internet usage. These findings suggest that parent education should be included as component of an anti-bullying program.

REFERENCES

Agatston, P., Kowalski, R., & Limber, S. (2007). Students' perspectives on cyber bullying. *The Journal of Adolescent Health*, *41*(6), S59–S60. doi:10.1016/j.jadohealth.2007.09.003 PMID:18047946

American Foundation for Suicide Prevention. (2017). *About suicide*. Retrieved July 23, 2017, from https://afsp.org/about-suicide/suicide-statistics/

American Psychological Association Zero Tolerance Task Force. (2008). Are zero tolerance policies effective in the schools? An evidentiary review and recommendations. *The American Psychologist*, *63*(9), 852–862. doi:10.1037/0003-066X.63.9.852 PMID:19086747

Berson, I., Berson, M., & Ferron, J. (2002). Emerging risk of violence in the digital age: Lessons for educators from an online study of adolescent girls in the United States. *Journal of School Violence*, *1*(2), 15–33. doi:10.1300/J202v01n02_04

Blueprints for Healthy Youth Development. (n.d.). *Olweus Bullying Prevention Program*. Retrieved June 23, 2017, from http://www.blueprintsprograms.com/evaluation-abstract/olweus-bullying-prevention-program

Borowsky, I. W., Taliaferro, L. A., & McMorris, B. J. (2013). Suicidal thinking and behavior among youth involved in verbal and social bullying: Risk and protective factors. *Journal of Youth and Adolescence*, *53*, S1–S12. doi:10.1016/j.jadohealth.2012.10.280 PMID:23790200

Centers for Disease Control and Prevention. (2016). *Understanding bullying*. Retrieved May 12, 2017, from https://www.cdc.gov/violenceprevention/pdf/Bullying_Factsheet.pdf

Centers for Disease Control and Prevention. (n.d.). *Connectedness as a strategic direction for the prevention of suicidal behavior*. Retrieved June 10, 2107, from http://www.cdc.gov/violenceprevention/pdf/suicide_strategic_direction-one-pager-a.pdf

Cook, C. R., Williams, K. R., Guerra, N. G., Kim, T. E., & Sadek, S. (2010). Predictors of bullying and victimization in childhood and adolescence: A meta-analytic investigation. *School Psychology Quarterly*, *25*(2), 65–83. doi:10.1037/a0020149

Davis, S., & Nixon, C. (2010). *The youth voice research project: Victimization and strategies*. Retrieved July 20, 2017, from http://njbullying.org/documents/YVPMarch2010.pdf

Durlak, J., & Weissberg, R. (2013). *Afterschool programs that follow evidence-based practices to promote social and emotional development are effective. Big views forward: A compendium on Expanded Learning*. Retrieved June 12, 2017, from http://www.expandinglearning.org/docs/Durlak&Weissberg_Final.pdf

Durlak, J., Weissberg, R., Dymnicki, A., Taylor, R., & Schellinger, K. (2011). The impact of enhancing students' social and emotional learning: A meta-analysis of school-based universal interventions. *Child Development*, *82*(1), 405–432. doi:10.1111/j.1467-8624.2010.01564.x PMID:21291449

Espelage, D., & Holt, M. (2013). Suicidal Ideation and School Bullying Experiences After Controlling for Depression and Delinquency. *The Journal of Adolescent Health*, *53*(1), S27–S31. doi:10.1016/j.jadohealth.2012.09.017 PMID:23790197

Farrington, D., & Ttofi, M. (2010). *School-based programs to reduce bullying and victimization*. Systematic review for The Campbell Collaboration Crime and Justice Group. Retrieved June 19, 2017, from www.ncjrs.gov/pdffiles1/nij/grants/229377.pdf

Finkelhor, D., Vanderminden, J., Turner, H., Shattuck, A., & Hamby, S. (2014). Youth exposure to violence prevention programs in a national sample. *Child Abuse & Neglect*, *38*(4), 677–686. doi:10.1016/j.chiabu.2014.01.010 PMID:24630440

Hawkins, D., Pepler, D., & Craig, W. (2001). Peer interventions in playground bullying. *Social Development*, *10*, 512–527. doi:10.1111/1467-9507.00178

Hazelden Foundation. (2016). *Violence prevention works: Olweus Bullying Prevention Program.* Retrieved July 1, 2017, from http://www.violencepreventionworks.org/public/bullying_effects.page

Heron, M. (2016). *Deaths: Leading causes for 2013.* National Vital Statistics Reports, 65(2). Hyattsville, MD: National Center for Health Statistics. Retrieved July 13, 2017, from https://www.cdc.gov/nchs/data/nvsr/nvsr65/nvsr65_02.pdf

Hertz, M., Donato, I., & Wright, J. (2013). Bullying and suicide: A public health approach. *The Journal of Adolescent Health*, *53*(10), S1–S3. doi:10.1016/j.jadohealth.2013.05.002 PMID:23790194

Institute of Medicine & National Research Council. (2014). *Building capacity to reduce bullying: Workshop summary.* Washington, DC: The National Academies Press.

Kosciw, J. G., Greytak, E. A., Giga, N. M., Villenas, C., & Danischewski, D. J. (2016). *The 2015 National School Climate Survey: The experiences of lesbian, gay, bisexual, transgender, and queer youth in our nation's schools.* New York: GLSEN.

Lee, T., & Cornell, D. (2009). Concurrent validity of the Olweus Bully/ Victim Questionnaire. *Journal of School Violence*, *9*(1), 56–73. doi:10.1080/15388220903185613

Limber, S. (2003). Efforts to address bullying in U.S. schools. *American Journal of Health Education*, *34*(5), S-23–S-29. doi:10.1080/19325037.2003.10603589

Mentors in Violence Protection. (2017). *MVP strategies.* Retrieved July 25, 2017, from http://www.mvpstrat.com/mvp-programs/high-school/

Mossberger, K. (2000). *The politics of ideas and the spread of enterprise zones.* Washington, DC: Georgetown University Press.

Musu-Gillette, L., Zhang, A., Wang, K., Zhang, J., & Oudekerk, B. (2017). *Indicators of school crime and safety: 2016 (NCES 2017-064/NCJ 250650)*. Washington, DC: U.S. Department of Justice.

National Academies of Sciences, Engineering, and Medicine. (2016). *Preventing bullying through science, policy, and practice*. Washington, DC: The National Academies Press. doi: 10.17226/23482

Newman, K., Fox, C., Roth, W., Mehta, J., & Harding, D. (2004). *Rampage: The social roots of school shootings*. New York: Basic Books.

Olweus, D. (1993). *Bullying at school: What we know and what we can do*. Malden, MA: Blackwell Publishing Ltd.

Olweus, D. (1996). *The revised Olweus Bully/Victim Questionnaire for Students*. Bergen, Norway: University of Bergen.

Olweus, D. (2013). School bullying: Development and some important challenges. *Annual Review of Clinical Psychology*, *9*(1), 751–780. doi:10.1146/annurev-clinpsy-050212-185516 PMID:23297789

Olweus, D., Limber, S., & Mihalic, S. F. (1999). Bullying Prevention Program: Blueprints for Violence Prevention. Boulder, CO: Center for the Study and Prevention of Violence, Institute of Behavioral Science, University of Colorado.

Perren, J., Corcoran, L., Cowie, H., Dehue, F., Garcia, D., McGuckin, C., ... Völlink, T. (2012). Tackling cyberbullying: Review of empirical evidence regarding successful responses by student, parents, and schools. *International Journal of Conflict and Violence*, *6*(2), 283–293.

Polanin, J., Espelage, D., & Pigott, T. (2012). A meta-analysis of school-based bullying prevention programs' effects on bystander intervention behavior. *School Psychology Review*, *41*, 47–65.

Seeley, K., Tombari, M. L., Bennett, L. J., & Dunkle, J. B. (2011). *Bullying in Schools: An Overview. Juvenile Justice Bulletin*. Office of Juvenile Justice and Delinquency Prevention.

Skiba, R., Reynolds, C. R., Graham, S., Sheras, P., Close Conoley, J., & Garcia-Vazquez, E. (2006). *Are Zero Tolerance Policies Effective in the Schools? An Evidentiary Review and Recommendations*. Zero Tolerance Task Force, American Psychological Association. Retrieved July 27, 2017, from http://www.apa.org/pubs/info/reports/zero-tolerance-report.pdf

Smith, J., Schneider, B., Smith, P., & Ananiadou, K. (2004). The effectiveness of whole school anti bullying programs: A synthesis of evaluation research. *School Psychology Review*, *23*(4), 547–560.

Strom, I., Thoresen, S., Wentzel-Larsen, T., & Dyb, G. (2013). Violence, bullying and academic achievement: A study of 15-year-old adolescents and their school environment. *Child Abuse & Neglect*, *37*(4), 243–251. doi:10.1016/j.chiabu.2012.10.010 PMID:23298822

Substance Abuse & Mental Health Services Administration. (2014). Bullying: Dispelling myths, enhancing prevention. *SAMHSA News, 22*(1). Retrieved July 26, 2017, from http://archive.samhsa.gov/samhsaNewsLetter/Volume_22_Number_1/bullying-myths-prevention.aspx#.WXoXiYjyuUk

Ttofi, M., & Farrington, D. (2008). Bullying: Short-Term and Long-Term Effects, and the Importance of Defiance Theory in Explanation and Prevention. *Victims & Offenders*, *3*(2/3), 289–312. doi:10.1080/15564880802143397

Ttofi, M., & Farrington, D. (2012). Bullying prevention programs: The importance of peer intervention, disciplinary methods and age variations. *Journal of Experimental Criminology*, *8*(4), 443–462. doi:10.100711292-012-9161-0

Ttofi, M., Farrington, D., & Baldry, A. (2008). *Effectiveness of programs to reduce school bullying: A systematic review*. Stockholm: Swedish National Council for Crime Prevention.

Vossekuil, B., Fein, R., Reddy, M., Borum, R., & Modzeleski, W. (2014). *The Final Report and Findings of the Safe School Initiative: Implications for the Prevention of School Attacks in the United States.* United States Secret Service and United States Department of Education. Retrieved July 20, 2017, from https://www2.ed.gov/admins/lead/safety/preventingattacksreport.pdf

Vreeman, R., & Carroll, A. (2007). A systematic review of school-based interventions to prevent bullying. *Archives of Pediatrics & Adolescent Medicine*, *161*(1), 78–88. doi:10.1001/archpedi.161.1.78 PMID:17199071

Chapter 5
Explaining Policy Adoption:
An Empirical Analysis

ABSTRACT

The response by government officials to heartbreaking events such as the suicides of Ryan Halligan and Phoebe Price has been mixed. At the state level, actions have ranged from attempting to use traditional aggression policies to halt the rise of cyber aggression crimes to passing new laws aimed specifically at cyber aggression. What explains the differences in the state responses? In exploring this question, this chapter examines the influence of state-level variables on policy adoption for four different forms of cyber aggression. For laws addressing cyberbullying, cyber-harassment, and cyberstalking, this chapter explores the level of legislative action concerning the updating and/or passing of new laws for the years 2007 through 2015. Pooled cross-sectional time series data that controls for variation between states and over time is used. Revenge porn laws are examined, but because they are relatively new, a cross-sectional analysis is presented for the year 2016.

INTRODUCTION

Chapters 3 and 4 discuss existing cyber aggression policies including their legislative history and preliminary policy evaluation of their effectiveness. Among the findings of these two chapters is that policy adoption for this issue area is that policy is far from uniform. An important question is, "why isn't there greater consistency among jurisdictions in the U.S. for addressing forms

DOI: 10.4018/978-1-5225-5285-7.ch005

of cyber aggression?" Although there are many factors that influence policy adoption, in this issue area public opinion has become and important factor in placing the topic on the agenda. In 1999, the tragedy at Columbine drew national attention to the problem of bullying. Similarly, in 2003, the suicide of Ryan Halligan became one of the first tragedies to alert the American public to the problem of cyberbullying and was instrument in the adoption of early state-level cyberbullying laws. The suicide of 15-year-old Phoebe Price in 2010 further increased calls by citizens for state-level school policies to address cyberbullying. As the Internet has evolved, so has the opportunities for online crimes and continued need to change existing policy. For example, the suicide of Amanda Todd in 2012, focused a spotlight on revenge porn or "nonconsensual porn." Existing policy was not designed to address this type of cybercrime and necessitated changes to current policy. Tyler Clementi's suicide in 2010 made us aware for the need for colleges and universities to provide a supportive environment for LGBTQ students on college campuses. These examples help illustrate the evolution of cyber aggression and how the circumstances surrounding these individual cases helped draw our attention to several barriers in crafting laws that can adequately address them. One barrier is that cybercrimes do not respect geographical boundaries. It can be difficult to pass a policy that will stop cybercrimes when culprits are beyond legal jurisdiction. In the case of Amanda Todd, the perpetrator was arrested and sentenced in the Netherlands, but jurisdictional issues can make it difficult to obtain justice (Nobullying.com, 2017).

The second lesson is that public demand (while important in placing an issue on the agenda) does not always result in policy change. Although Amanda Todd's death brought public demand for legislation addressing online bullying; Bill C-13 was proposed in Canadian Parliament to help protect children from cyberbullying but never passed. Similarly, following the suicide of Tyler Clementi in 2010 there were calls for government action. In response, the federal government has made several attempts to pass the Tyler Clementi Higher Education Anti-Harassment Act which would require institutions of higher education receiving federal funds to have an anti-harassment policy in place. It was introduced on March 27, 2014 and reintroduced on March 18, 2015 with Congress failing to enact it on either occasion. It has again been reintroduced on April 27, 2017 but it seems unlikely that it will be enacted (Civic Impulse, 2017). In addition, although all 50 states have adopted anti-bullying policy, some of the policies are only symbolic. In some states, public policy defines bullying but do not prescribe a solution or mandate action. Finally, even though many states have adopted criminal sanctions for acts of cyber aggression, punishment vary widely. In some states, bullying may be a felony and in others a misdemeanor, even in the instance of suicide. The teens that were prosecuted for Phoebe Prices suicide were sentenced to community service while Tyler Clementi's roommate spent 20 days in jail (Leefeldt, 2016). Why are there differences among states in their response to public demand for government action? This chapter explores this topic beginning with a summary of the literature on policy adoption.

STATE-LEVEL POLICY ADOPTION LITERATURE

As discussed in Chapters 3 and 4, the legal response to online aggression varies significantly based on jurisdiction. All 50 states, territories and the federal government now outlaw stalking and harassment, but not all have updated their laws to include their online counterparts (i.e. cyberstalking, and cyber-harassment). Although the federal government has yet to adopt policy outlawing bullying or cyberbullying, the states have been experimenting with legal solutions. Following the Columbine shootings, Georgia became the first state to enact anti- bullying policy in 1999 with Montana became the last state to adopt an anti-bullying policy in 2015 (Cyberbullying Research Center, 2016a). These laws are far from consistent. Although all 50 states have anti-bullying laws, only 23 address cyberbullying while 48 include language concerning electronic harassment and 14 include sanctions for off campus behavior (Cyberbullying Research Center, 2016a). These laws continue to expand as cybercrimes evolve with new forms of harassment including revenge porn. States have been struggling to keep up as social media changes; currently 38 states and DC have adopted revenge porn laws while only 26 states have updated school anti-bullying policy to include revenge porn (Cyberbullying Research Center, 2016b; Cyber Civil Rights Initiative, 2017).

What explains the difference in how states are responding to acts of online aggression? The literature on policy adoption suggests a number of factors. Early research indicates that socio-economic factors including state wealth was important to policy adoption (Walker, 1969; Gray, 1973; Hwang & Gray, 1991). Other studies (Meier, 1994; Mooney & Lee, 1995) suggest that both politics and public demand or need increase the change of policy adoption.

Additionally, Nicholson-Crotty (2009) found that salience was associated with a policy innovations spreading quicker from one state to another. Whether or not an issue is even taken up by policymakers depends on characteristics of the problem. According to Rochefort and Cobb (1994) five attributes of a problem will help influence public opinion and determine whether policymakers will turn their attention to it: causality, severity, incident, proximity and crisis. Causality refers to who is believed to be at fault. Does the problem exist because of individual actions or is society or government to blame? Severity concerns how serious the problem is while incident reflects how often the problem occurs. Proximity indicates how closely individuals feel affected by an issue while crisis refers to whether the media or policymakers are using the word crisis to describe the problem. Although each of these factors influences policy adoption, their relative importance depends on the policy area being examined. Policy that concerns the criminalization of online aggression activities could be considered regulatory. Although regulatory policies tend to be dominated by political factors, politics may play less of a role when dealing with the criminalization of personal behavior. In these instances, the main political issue concerns whether the public believes punishment is appropriate for the crime being committed (Lowi, 1964; Ripley & Franklin, 1980). The literature on criminal justice policy finds that a number of political factors including partisanship, ideology, public opinion, media framing of crime related stories and race relations drive policy decisions (Carmines & Stimson, 1989; Mooney & Lee, 2000; Kellstedt, 2003).

The literature suggests a number of variables that can be used to explain variation in state policy. Because cyber aggression continues to evolve with changes in telecommunication technology, this study analyzes the impact of these factors overtime. For laws addressing cyberbullying, cyber-harassment and cyberstalking, this chapter explores the level of legislative action to update and/or pass new laws for 2007 through 2015. Pooled cross-sectional time series data that controls for variation between states and over time is used. This study begins with the year 2007 because few states had adopted laws to address cyber aggression prior to this year. It was not until media coverage of events, such as the suicide of Megan Meier in 2006, did policymakers at the state level begin to adopt laws that address these crimes. Revenge porn laws will also be examined but because they are relatively new, a cross-sectional analysis will be presented for the year 2016.

EMPIRICAL METHODS: DATA AND METHODS

Cyberbullying Laws

The section represents an extension of prior research (e.g. McNeal, Kunkle, & Bryan, 2016) that examined which factors helped to predict which state would be more likely to updated existing bullying laws to include cyberbullying. This previous research attempted to explain the existence of cyberbullying policy but did not take into consideration the differences among the policies adopted by each state. In 2015, Montana became the last state to adopt a bullying policy and practically speaking, it was strictly symbolic. The law provided a definition of bullying but did not mandate any actions to be taken. In contrast, other states (e.g. Florida, Connecticut and Arkansas) have updated their traditional bullying laws to include cyberbullying and online harassment, and impose both criminal and school sanctions. McNeal, Kunkle, and Bryan (2016) found that resources played the biggest role in whether states updated existing anti-bullying laws to include cyberbullying; wealthier states were more likely to take the lead in this policy area. Does this hold true when the provisions of the law are considered? To examine this question, this chapter tests the impact of independent and control variables suggested by the literature on policy adoption and diffusion on the number of provisions included in each state's cyberbullying policy. The dependent variable represents a count from 0 to 5 where 1 point is given for the inclusion of the following provision in a state's school anti-bullying policy: direct mention of cyberbullying or online harassment; criminal sanctions for bullying; school sanctions for bullying; requirement for school policies on bullying; and inclusion of off-campus activities. This variable was created using information from the Cyberbullying Research Center (2016a). Independent and control variables include measures for politics, state resources and citizen demands/ needs.

Policy adoption is a political activity where institutional actors must decide which policy solution should be used to solve a public problem. Depending on the policy area, any number of political actors including political parties, interests groups, constituents and governments bureaucracies may try to influence which policy solution is selected (Anderson, 2011). To control for the possible influence of state-level interests groups, a measure for the overall impact of interest group on the state political system was included. This five-point scale indicates the overall impact that interest groups have in a state political system as related to other groups. A score of 5 indicates that interest groups are dominating in the political system as compared to other groups, and a score of 0 indicates that they are subservient (Thomas & Hrebenar, 2008). Both elected official ideology and citizen ideology may influence policy adoption. The passage of stricter legislation to control criminal behavior is associated with a conservative ideology. It is expected that in states where state legislatures and citizens are more conservative, existing laws would be more likely to contain stricter provisions. In order to control for the ideology of citizens and elected officials, two indices were added each ranging from 0 to 100 with higher scores indicating greater liberalism (Fording, 2012). These indices are updated versions of those first proposed by Berry et al. (1998). The measure of citizen ideology is based on interest group ratings of members of Congress of each state and estimated scores of challengers and election results (Berry et al., 1998, pp. 330-331). The measure of government ideology is also based on interest group ratings and is computed using weighted ideological score averages for the state governor and both major parties in each legislative chamber (Berry et al., 1998, 332-333).

Another factor that might influence the number of provisions included in a state's cyberbullying policy is the presence of women in the state legislature. Because females are more likely to be victims of acts of aggression (Baum, Catalano, Rand, & Rose, 2009; Black, et al, 2011), it is predicted that more women in the state legislature would encourage the updating of existing laws. The presence of women legislators is measured using the percent women in a state legislature for each year (Center for American Women and Politics, 2016). Whether or not a state has term limits may also influence how legislators vote. There is debate however on how term limits might influence policy decisions. Supporters of legislative term limits argue that they could make elections more competitive and help to unseat career politicians who are unresponsive to their constituents (Will, 1993) while detractors (Glazer & Wattenberg, 1996) argue that the desire to get reelected makes representatives more receptive to their constituents' concerns. If this argument is true, imposing terms limits could make politicians less responsive to the public. The presence of legislative term limits is measured using a variable coded 1 if the state has legislative term limits and 0 otherwise has been included (National Council of State Legislatures [NCSL], 2005). Similarly, a measure was added for gubernatorial term limits. Gubernatorial election year was included as a final political factor coded 1 if it was a gubernatorial election year and 0 otherwise. Unlike most state-level elections, gubernatorial races garner significant public and media attention (Partin, 2001; Freedman & Fico, 2004). Given the high stakes nature of these elections, a siting governor may attempt to push for legislation popular with voters during an election year. Government agencies are also important stakeholders that can influence impact policy adoption. It may be easier to adopt policy if a state has greater control over the resources of the implementing agency. As an indicator of control over state agencies, an index measuring state legislative oversight of the bureaucracy was included. It was constructed using four-point scale where 1 indicates that the state legislature has no oversight power of the bureaucracy and 4 indicates that the legislature can impose costs and/or suspend rules (Gerber, Maestas & Dometruis, 2005).

Several measures of state resources were also added including urbanization, educational attainment and gross state product. Gross state product is measured over time in millions of dollars (Bureau of Economic Analysis, 2014) while urbanization is measured by the percent of the population living in urban areas (United States Census Bureau, 2012a, 2012b). Educational attainment is measured over time by the percent of the state population age 25 or older with a college degree with data from the U.S. Census Bureau (2012a). Finally, public need/demand (Meier, 1994; Mooney & Lee, 1995) has been found to be an important factor in predicting policy adoption. The first measure of need is percentage of households with Internet access within the state (National Telecommunications and Information Administration [NTIA], 2014). It is expected that in states with greater Internet usage, the likelihood of cybercrimes would be greater. Because cyberbullying is directed at children and teens, a measure of the percent of the state's population under the age of 18 was included (U.S. Census Bureau, 2012a). As a final indicator of need, the number of complaints regarding Internet crimes per 100,000 people were included (Internet Crime Complaints Center, various years). It is expected that in states with a greater occurrence of cybercrimes, there will be a greater likelihood of updating existing bullying policies to include a greater number of provisions to reduce the incidences of cyberbullying. Finally, measures were added to control for regional variations in the occurrence of cyberbullying threats. A 2011 survey found that cyberbully is more likely to take place in the Midwest. The survey found that 10.2 percent of students in the Midwest, 9.2 percent of the students in the Northeast, 8.8 percent of the students in the students in the South and 8.4 percent of the students in the West reported being the victim of cyberbullying either in school or off campus (U.S. Department of Education, 2013, p. T-23). To control for regional variations dummy variables were included for the Midwest, South and West; the Northeast serves as the reference group.

Cyberstalking and Cyber-Harassment Laws

Dependent variables were created for both state-level cyberstalking and cyber-harassment laws. Both dependent variables were coded 1 if the state had updated existing laws to include electronic communication and 0 otherwise. A number of sources were used to create the dependent variables (NCSL, 2015; Hazelwood & Koon-Magnin, 2013; Goodno, 2007; Fukuchi, 2011; Cyberbullying Research Center, 2016). These models control for the same list of independent and control variables as those used in the model for the extent of cyberbullying policies with the exception of replacing the measure for the percentage of the state population under the age of 18 with the percent of population over the age 65 (U.S. Census Bureau, 2012a). The change in age variables is based on the legal definition of these forms of cyber aggression. Cyberbullying specifically concerns aggressive behavior among children and adolescences (U.S. Department of Health & Human Services, 2015). It is expected that states with an older population, in theory, would have less demand for changes in existing laws to stalking and harassment because they might be less likely to perceive them as a crime. Laws regulating stalking were not passed until the 1990s and an older generation may be more likely to consider this issue a personal matter and not a criminal issue. In addition, individuals over the age of 65 have been found to be among those less likely to use the Internet (NTIA, 2014). Limited Internet usage may also influence opinion regarding the need to strengthen existing laws to include online forms of aggression.

Revenge Porn Laws

A dependent variable was created for the severity of state-level revenge porn laws coded 2 for a felony offence, 1 for a misdemeanor and 0 if the state did not have a revenge porn law. This variable was created using information from the Cyber Civil Rights Initiative (2017). Unlike the models for cyberbullying, cyberstalking and cyber-harassment laws; this model represents a cross sectional analysis for the year 2016. Independent and control variables include politics, state resources and citizen demands/ needs. Similar to the other models, variables were included for overall interest group impact, ideology of elected officials and citizens, legislative term limits, percent women in the state legislature, percent urban, education attainment, Internet access and Internet crime rate. In addition, because of the link between sexting and revenge porn laws, a variable was added coded 1 if the state had a sexting law and 0 otherwise (Cyberbullying Research Center, 2016b). It is expected that states with sexting laws would have stronger revenge porn laws. Because the sample size is smaller for this model, multicollinearity was a greater a concern than for the other models. Therefore, fewer independent and control variables were included.

FINDINGS AND DISCUSSION

Cyberbullying Laws

In Table 1, the dependent variable is coded so that higher scores are associated with a greater number of provisions contained in existing cyberbullying laws. Since the dependent variable is measured over time pooling the fifty states and the measure represents a count, Poisson panel regression is used. The findings in Table 1 are not consistent with earlier research. While McNeal, Kunkle, and Bryan (2016) found that state resources played a dominate role in updating existing anti-bullying laws to include cyberbullying, the number of provision included in current cyberbullying laws is not dominated by any specific factor. In addition, the findings are not consistent with the literature on criminal justice policy which points to several political factors in the adoption of laws in this area. Although policy adoption is political, with various stakeholders attempting to influence which proposal will be implemented, political factors were found to have the least influence on the provisions in existing cyberbullying policies. The only political constraint found to be significant was the ideology of citizens. The findings indicate that states dominated with conservative citizens were more likely to strengthen and expand anti-bullying laws. This finding is consistent with the literature. It was assumed that conservative citizens would be more likely to support the passage of stricter legislation to control criminal behavior. This is predicated on the belief that cyberbullying laws would be portrayed in the media in terms of law and order policies. The results of this study may be different from previous studies because it focuses on how extensive the policy is and not simply if a policy was adopted.

As expected, gross state product was significant and positively related to the number of provisions adopted. This is consistent with early research that found state wealth was an important determinate of policy adoption

Table 1. State cyberbullying laws, 2007-2015

Variables	The Extent of State-Level Cyberbullying Policy	
	β (se)	**p>lzl**
Interest Group Strength		
Overall interest group impact ^{i,t}	.016(.130)	.189
Political Constraints		
Ideology of elected officials ^{i,t}	- .004(.003)	.141
Ideology of citizens ^{i,t}	**-.016(.007)**	**.019**
Legislative term limits ^{i,t}	-.314(.194)	.106
Gubernatorial term limits ^{i,t}	-.107(.236)	.651
Gubernatorial election ^{i,t}	-.092(.070)	.189
% women in the state legislature ^{i,t}	2.81e-4(.013)	.987
State legislative oversight of bureaucracy ^{i,t}	-.028(.067)	.699
State Resources		
Gross state product ^{i,t}	**3.57e-7(2.09e-7)**	**.087**
Urban population (%)^{i,t}	1.47e-4(.008)	.985
Education attainment ^{i,t}	**-.047(.028)**	**.092**
Demands/Needs		
Percent citizens 18 or younger ^{i,t}	**-.159(.048)**	**.001**
Internet Access ^{i,t}	**.081(.014)**	**.012**
Internet Crime Rate ^{i,t}	-.001(.001)	.415
Midwest	-.239(.273)	.382
South	.125(.314)	.690
West	-.249(.319)	.436
Constant	1.708(1.719)	.320
Wald Chi^2 (17)	173.45	.0000
Number of Panels	50	
N	450	

Note: Poisson panel regression data for the 50 states. Unstandardized Poisson coefficients are presented with standard errors in parenthesis. Subscript *i* contains the unit to which the observations belong, in this case the state, and controls for variation in state legislative activity between the states. Subscript *t* represents the time or year the variable was measured. Statistically significant coefficients at .10 or less in bold.

(Walker, 1969; Gray, 1973; Hwang & Gray, 1991). Percent urban was not found to be significant while education attainment was significantly related to policy adoption; it was not in the expected direction. It is possible that in this model, educational attainment is measuring need instead of an indicator of a state resource. Like the measures of state resources, the results for needs/demand were mixed. The finding that in states where there is greater Internet access, there is a greater likelihood that these policies will be adopted is in the expected direction. In states where there is greater access to the Internet, it would be expected that there would be greater incidences of cyberbullying. This supports the literature on need/demand (Meier, 1994; Mooney & Lee, 1995) as well as Rochefort and Cobb's (1994) argument that problem characteristics influence public opinion.

The findings for age were significant but not in the expected direction. States with a higher percentage of the population under the age of 18 in a state were actually less likely to have a greater number of policy provisions to address cyberbullying. This is not entirely unexpected. McNeal, Kunkle, and Bryan (2016) found that the percentage of population under 18 was not a significant predictor of whether a state had a cyberbullying law. They argued that Internet access was a better predictor of whether cyberbullying was a problem in the state and the findings from this study support this argument. Finally, dummy variables for region and number of Internet crimes were found unrelated to the extent of policy adoption. The findings for the measure of Internet crimes was unrelated to the extent of policy adoption might be because it is a general measure of Internet crimes ranging from identity theft to romance scams. Overall, the results suggest that need is not the driving force in determining the number of provisions included in a state's cyberbullying laws.

Cyberstalking and Cyber-Harassment Laws

In Table 2, the dependent variables are coded so that higher scores associated with increased likelihood that a state will modify existing stalking or harassment legislation to include electronic communication. Since the dependent variables are measured over time pooling the fifty states and the dependent variables are binary, a binary panel logistic regression is used. The findings in Table 2 suggest that circumstances that lead to the updating of earlier stalking laws differ from those that lead to the modification of harassment laws. Specifically, the findings suggest that a limited number of factors influence the updating of existing stalking laws while several factors were found related to the modification of existing harassment policy. This is surprising because the definition of cyber-harassment is closer to that of cyberbullying (with the main difference being the age of the victim). Both cyber-harassment and cyberbullying concern the use of the Internet to torment the victim while cyberstalking indicates that there is an immediate credible threat to an individual (NCSL, 2015). While it is expected that cyberstalking would garner greater legislative attention, it is possible that greater media coverage of online harassment has resulted in more public attention.

The updating of both stalking and harassment laws to include online activities were found to be influenced by political factors as predicted by the literature on criminal justice policy. The political factors relevant to the adoption of laws updating existing harassment laws include interest group strength, legislative term limits and whether it is a gubernatorial election year; each of these variables is negatively associated with updating current laws. States are more likely to add cyber activities to existing harassment laws when there are no legislative term limits, state-level interest group are weak and it was not a gubernatorial election years. On the other hand, states were more likely to strengthen their stalking laws when the citizens are more conservative, it was not a gubernatorial election year and there are term limits on governors. The only outcome shared by both models is the finding that both types of laws are less likely to be adopted during a year when governors are being elected. It is possible that during an election year, the sitting governor may be unwilling to take up any issue that could be controversial. The finding that harassment laws are less likely to updated when there are term limits on the state legislature and stalking laws are more likely to be updated when there are term limits are placed on governors suggest that state legislatures are taking the lead in this policy area. Taken together these findings suggest that action is more likely to be taken when the power of the governor is weaker in comparison to the state legislature.

Like the model in Table 1, the percentage of women in the state legislature and state legislature control over state agencies were found to be unrelated to policy adoption of cyber aggression laws. These findings are not consistent with the literature. Agency administrators are often among the actors taking the lead to update existing laws, but they may not be as involved in this policy area. It was expected that the percentage of women in the state legislature would be positively related to the adoption of policy to address cyber aggression. It is possible that that the percentage of women in the state legislature is acting as a measure of the balance in power between the state legislature and the governor. Not all states have the same probability that women will run for state office. Research (King, 2002) finds that states that have a greater percentage of the women in the state legislature tend to be those where the legislators receive lower pay and have shorter sessions. This suggests that these are states with weaker state legislatures. This finding supports others

Table 2. State cyberstalking and cyber-harassment laws, 2007-2015

Variables	Presence of Regulations Cyberstalking		Presence of Regulations Cyber-Harassment	
	β (se)	p>|z|	β (se)	p>|z|
Interest Group Strength				
Overall interest group impact ^{i,t}	-2.778(1.903)	.144	**-4.315(1.985)**	**.030**
Political Constraints				
Ideology of elected officials ^{i,t}	-.007(.028)	.776	.028(.038)	.458
Ideology of citizens ^{i,t}	**-.118(.065)**	**.071**	-.048(.082)	.562
Legislative term limits ^{i,t}	-.717(2.525)	.777	**-6.086(2.578)**	**.018**
Gubernatorial term limits ^{i,t}	**5.236(3.160)**	**.098**	1.227(3.342)	.714
Gubernatorial election ^{i,t}	**-1.289(.695)**	**.064**	**-1.416(.857)**	**.098**
% Women in the state legislature ^{i,t}	.103(.125)	.411	-.057(.155)	.712
State legislative oversight of bureaucracy ^{i,t}	.172(.716)	.810	1.107(.860)	.198
State Resources				
Gross state product ^{i,t}	3.68e-6(3.09e-6)	.234	**9.89e-6(3.14e-6)**	**.002**
Urban population ^{i,t}	.040(.118)	.736	.032(.112)	.777
Education attainment (%) ^{i,t}	-.160(.350)	.648	-.014(.361)	.969
Demands/Needs				
Percent citizens 65 or older ^{i,t}	**1.320(.676)**	**.051**	**2.529(.902)**	**.005**
Internet Access ^{i,t}	**.274(.121)**	**.024**	**.376 (.147)**	**.011**
Internet Crime Rate ^{i,t}	.006(.019)	.749	.020(.015)	.178
Midwest	-2.751(4.033)	.495	**6.347(3.506)**	**.070**
South	-1.473(4.943)	.765	**12.353 (4.914)**	**.012**
West	.883(4.424)	.842	3.151(4.623)	.495
Constant	-18.608(14.148)	.188	**-50.508(17.343)**	**.004**
Wald Chi^2 (17)	33.06	.0111	41.72	.0007
Number of Panels	50		50	
N	450		450	

Note: Binary panel logistic regression data for the 50 states. Unstandardized logistic coefficients are presented with standard errors in parenthesis. Subscript i contains the unit to which the observations belong, in this case the state, and controls for variation in state legislative activity between the states. Subscript t represents the time or year the variable was measured. Statistically significant coefficients at .10 or less in bold.

that indicate that cyber aggression laws are more likely to be adopted when the state legislature is more powerful in comparison to the governor.

Consistent with the findings from Table 1, state resources were unrelated to policy adoption of cyber aggression policy. The only significant relationship was between the adoption of cyber-harassment policy and gross state product. Need or demand played a greater role in policy adoption than state resources. As predicted, states where a greater percentage of the population had Internet access were more likely to adopt both online harassment and stalking policies. Not expected is the finding that a higher percentage of citizens over 65 was found to be positively associated with both cyber-harassment and cyberstalking laws. One explanation may be how harassment is perceived by the population. Although individuals 65 and older are among those less likely to use the Internet (NTIA, 2014), their perception about the likelihood of cyber aggression may be higher than other segments of the population even if they are less likely to be victims of these crimes. Finally, regional differences were found in the adoption of cyber-harassment laws. States in the South and Midwest were more likely to update current harassment laws than states in the Northeast. Regional differences may be based on relative differences between the power of the legislature and governor or the ideology of the citizens and lawmakers.

Revenge Porn Laws

In Table 3, the dependent variable is coded so that higher scores associated with a higher legal penalty for committing a crime of revenge porn. Since the dependent variable is ordinal, an ordered logistic regression model was used. Revenge porn is a form of harassment / bullying and would be expected to have similar results to the models for cyberbullying and cyber harassment. Over all, the findings for level of penalty for revenge porn was similar to those for the number of provisions in state school cyberbullying laws in that only a limited number of factors were relevant in explaining differences in state policies for these two crimes. The dependent variables for both of these models are looking it differences in state policies, and not simply if the state has adopted a policy. This suggests that different forces are at play when the legislature is considering what policy to implement compared to the debate over whether adopt a policy at all. One of those forces might be the stakeholders; individuals involved in getting an issue on the agenda might not be the same who try to influence which policy is adopted. The actors that

Table 3. State revenge porn laws, 2016

Variables	The Extent of State-Level Revenge Porn Policy	
	β (se)	p>|z|
Interest Group Strength		
Overall interest group impact	**.810(.475)**	**.088**
Political Constraints		
Ideology of elected officials	.019(.015)	.219
Ideology of citizens	- .016(.039)	.676
Legislative term limits	-. 434(.719)	.546
% women in the state legislature	**.142(.070)**	**.044**
State Resources		
State Sexting Laws	1.053(.760)	.166
Urban population (%)	-.018(.027)	.521
Education attainment	**-.208(.121)**	**.085**
Demands/Needs		
Internet Access	**.189(.082)**	**.021**
Internet Crime Rate	-.006(.007)	.381
LR Chi2 (10)	18.91	**.0414**
Pseudo R^2	.1912	
N	50	

Note: Ordered logistic regression estimates with standard errors in parentheses. Reported probabilities are based on two-tailed tests. Statistically significant coefficients at .10 or less in bold.

played a role in the determination of the legal penalty for revenge porn were unique to this policy area. Unlike the other policy areas examined in this chapter, the most important stakeholders were female state legislature and interest groups. States with a greater percentage of state legislators who are female and strong interest groups were more likely to adopt policy classifying revenge porn as a felony. The percentage of state legislators who are female was not a significant variable in any of the other models. It was thought that this might because state with more female state legislators also tends to be states with more powerful governors. This may still be true. How revenge porn is viewed as a crime might explain why having more females in state legislature might influence policy adoption for this policy. Females seem to be the most stigmatized by porn either as victims or willing participants; it seems that male porn does not carry with it the same degree or type of stigma. While male strippers are considered hot; female strippers are viewed as having moral defects.

The only other variables found to be significant predictors of the criminal penalty for revenge porn in a state were education attainment and Internet access. Education attainment was found to be negatively related to the severity of the punishment adopted for revenge porn. Although education attainment was include as an indicator of wealth, it is possible that in this model it is acting as a measure of need. The only variable found to be significant across the four policy areas in this chapter was that the extent to Internet access. This suggests that the greater incidences of a cybercrime helped lead to the perception that it was a problem that government should address as well as influences the type of policy adopted.

FUTURE RESEARCH DIRECTIONS

As with the need for greater research on all forms of cyber aggression including the causes of intimate partner violence and strategies to disrupt the cycle of abusive relationships, there is a need for more extensive research on why states adopt different policies to address these issues. This chapter represents a preliminary attempt at addressing the latter question. The findings suggests that laws meant to update existing stalking laws follow the pattern of adoption predicted in the literature on both regulatory and criminal justice policy with political factors taking the lead in predicting adoption. On the other hand, updating existing harassment policy was found to be explained by a divergent set of factors including politics and citizen demand or need. The findings for the number of provisions in state school cyberbullying laws and the level of penalty for revenge porn suggested only a limited number of factors were relevant in explaining differences in state policies for these two crimes. The outcomes of this research hints at the possibility that these criminal offenses have been framed differently in public debate but there is not sufficient evidence to support or disprove this argument. Future research is needed to examine how the language used in debate over the adoption of these policies differed and the impact this had on adoption. In addition, future consideration is needed to examine why the policies adopted to curb cyberstalking is like those used to address more traditional criminal offenses while the same is not true for the other three forms of cyber aggression studied in this chapter.

Furthermore, this chapter used unprecise instruments to measure the dependent variables for cyberstalking and cyber-harassment. They were coded 1 if a state had updated existing laws to include electronic communication. Although this research differentiated among the states in terms of state-level provisions for cyberbullying and revenge porn, this study did indicate the level of evidence needed to prove a crime had taken place (intent) for any of the cybercrimes examined. With the exception of the model examining revenge porn laws, there was no attempt to measure differences among states with regard to the level of punishment prescribed for these offences. While it is important to determine which states have updated or expanded existing laws, it is also essential to evaluate which of these new laws are successful in reducing the occurrence of these cybercrimes. Additionally, this chapter examined a time period in which the Internet has become widespread and the problem of cybercrimes has become extensive. Some states, such as California, began updating their existing laws early on, when Internet usage and cybercrimes were limited. Further research is needed to determine why some states acted as "bellwether" states, addressing these issues before they became prevalent. Finally, the model for revenge porn laws represents a preliminary examination of this policy area and unlike the other laws studied in this chapter; it relies on data from a single time period (a cross-sectional design). The findings from this model may suggest outcomes that are only relevant for a specific period of time. Future research of this policy area should be done over time using panel data.

CONCLUSION

This chapter started with the premise that although advances in telecommunication technology have opened up new avenues for communication and freedom of expression, it comes at a price. This new technology is also being used by those who choose it as a vehicle to exploit and victimize others. An important question is, "how has government acted to protect citizens from this new form of crime?" It focuses on four related cybercrimes (cyber-harassment, cyberstalking, cyberbullying, and revenge porn). Addressing these crimes is particularly difficult because they often go unreported. This study represents a preliminary effort that explores which states have taken the basic steps of updating existing bullying, stalking and harassment laws to include electronic communication. Because these laws can be regarded as a form of regulatory policy, it was expected that although state resources and citizens demands would play an important role in their adoption, political factors would dominate the adoption process. The findings were not entirely as expected. Although adoption of legislation for updating stalking laws followed existing theory on regulatory policy (Lowi, 1964; Ripley and Franklin, 1980), the updating of existing harassment, bullying and revenge porn laws did not.

Even though harassment, revenge porn and bullying are defined similarly to stalking (where the main difference among these crimes is a lack of an immediate credible threat), the factors that predicted the updating of their laws were significantly different. Political factors were found to dominate the process of updating existing stalking laws, while politics and citizen demand played a more equal role in the updating of existing harassment laws. The patterns for cyberbullying and revenge porn were less clear; no set of factors dominated. These findings suggest two things. The first is that closer attention to Rochefort and Cobb (1994) argument that problem attributes influences public opinion. One of the few commonalities across the four policy areas was that the extent to Internet access was an important predictor of state policy. This suggests that the greater incidences of a cybercrime helped lead to the perception that it was a problem that government should address. The second is that it is necessary to take a step back when examining policy adoption and consider the agenda setting process. During this part of the policy process, decisions are made concerning which issues will be taken up by the legislature. At this time, political actors work to define (frame) issues to bring about preferred outcomes. Although these crimes are legally similar, the findings imply that the perception of political actors is that these crimes are significantly different. This suggests that framing is important in both placing an issue on the agenda and the policy decision made after an issue has been taken up.

REFERENCES

Anderson, J. (2011). *Public policymaking: An introduction* (7th ed.). Boston: Cengage.

Baum, K., Catalano, S., Rose, K., & Rand, M. (2009). *Stalking Victimization in the United States* (NCJ 224527). Bureau of Justice Statistics Special Report. Washington, DC: U.S. Department of Justice. Retrieved May 9, 2014, from http://www.bjs.gov/index.cfm?ty=pbdetail&iid=365

Berry, W., Ringquist, E., Fording, R., & Hanson, R. (1998). Measuring citizen and government ideology in the American states, 1960-1993. *American Journal of Political Science*, *42*(1), 327–348. doi:10.2307/2991759

Black, M., Basile, K., Breiding, M., Smith, S., Walters, M., Merrick, M., . . . Stevens, M. (2011). *The National Intimate Partner and Sexual Violence Survey (NISVS): 2010 Summary Report*. Atlanta, GA: National Center for Injury Prevention and Control, Centers for Disease Control and Prevention. Retrieved July 25, 2014, from https://www.cdc.gov/violenceprevention/pdf/nisvs_report2010-a.pdf

Bureau of Economic Analysis. (2014). *Interactive tables*. Retrieved July 10, 2014, from http://bea.gov/iTable/index_nipa.cfm

Carmines, E., & Stimson, J. (1989). *Issue evolution: Race and the transformation of American politics*. Princeton, NJ: Princeton University Press.

Center for American Women and Politics. (2016). *Women in state legislatures*. Retrieved February 9, 2017, from http://www.cawp.rutgers.edu

Civic Impulse. (2017). *S. 954 — 115th Congress: Tyler Clementi Higher Education Anti-Harassment Act of 2017*. Retrieved December 23, 2017, from https://www.govtrack.us/congress/bills/115/s954

Cyber Civil Rights Initiative. (2017). *38 states + DC have revenge porn laws*. Retrieved March 25, 2017, from https://www.cybercivilrights.org/revenge-porn-laws/

Cyberbullying Research Center. (2016a). *Bullying laws across America*. Retrieved November 18, 2016, from http://cyberbullying.org/bullying-laws

Cyberbullying Research Center. (2016b). *Sexting laws across America*. Retrieved November 18, 2016, https://cyberbullying.org/sexting-laws

Fording, R. (2012). *State ideology data*. Retrieved June 15, 2014, from http://rcfording.wordpress.com/state-ideology-data/

Freedman, E., & Fico, F. (2004). Whither the experts? Newspaper use of horse race and issue experts in coverage of open governors' races in 2002. *Journalism & Mass Communication Quarterly*, *81*(3), 498–510. doi:10.1177/107769900408100303

Fukuchi, A. (2011). A balance of convenience: Burden-shifting devices in criminal cyberharassment law. *Boston College Law Review. Boston College. Law School*, *52*(1), 289–338.

Gerber, B., Maestas, C., & Dometrius, N. (2005). State legislative influence over agency rulemaking: The utility of ex ante review. *State Politics & Policy Quarterly*, *5*(1), 24–46. doi:10.1177/153244000500500102

Glazer, A., & Wattenberg, M. (1996). How will term limits affect legislative work? In B. Grofman (Ed.), *Legislative term limits: Public choice perspectives* (pp. 37–46). Boston: Kluwer. doi:10.1007/978-94-009-1812-2_3

Goodno, N. (2007). Cyberstalking, a new crime: Evaluating the effectiveness of current state and federal laws. *Missouri Law Review*, *72*(1), 125–197.

Gray, V. (1973). Innovations in the states: A diffusion study. *The American Political Science Review*, *67*(04), 1174–1185. doi:10.2307/1956539

Hart, B., & Klein, A. (2013). *Practical Implications of Current Intimate Partner Violence Research for Victim Advocates and Service Providers, document no. 244348*. Washington, DC: U.S. Department of Justice.

Hazelwood, S., & Koon-Magnin, S. (2013). Cyber stalking and cyber harassment legislation in the United States: A qualitative analysis. *International Journal of Cyber Criminology*, *7*(2), 155–168.

Holbrook, T., & La Raja, R. (2008). Parties and elections. In V. Gray & R. L. Hanson (Eds.), *Politics in the American States: A comparative analysis* (9th ed.). Washington, DC: Congressional Quarterly Press.

Internet Crime Complaint Center. (2006). *2006 Internet Crime Report*. Retrieved June 23, 2014, from http://www.ic3.gov/media/annualreports.aspx

Internet Crime Complaint Center. (2007). *2007 Internet Crime Report*. Retrieved June 23, 2014, from http://www.ic3.gov/media/annualreports.aspx

Internet Crime Complaint Center. (2008). *2008 Internet Crime Report*. Retrieved June 23, 2014, from http://www.ic3.gov/media/annualreports.aspx

Internet Crime Complaint Center. (2009). *2009 Internet Crime Report*. Retrieved June 23, 2014, from http://www.ic3.gov/media/annualreports.aspx

Internet Crime Complaint Center. (2010). *2010 Internet Crime Report*. Retrieved June 23, 2014, from http://www.ic3.gov/media/annualreports.aspx

Internet Crime Complaint Center. (2011). *2011 Internet Crime Report*. Retrieved June 23, 2014, from http://www.ic3.gov/media/annualreports.aspx

Internet Crime Complaint Center. (2012). *2012 Internet Crime Report*. Retrieved June 23, 2014, from http://www.ic3.gov/media/annualreports.aspx

Internet Crime Complaint Center. (2013). *2013 Internet Crime Report*. Retrieved June 23, 2014, from http://www.ic3.gov/media/annualreports.aspx

Kellstedt, P. (2003). *The mass media and the dynamics of American racial attitudes*. Cambridge, UK: Cambridge University Press. doi:10.1017/CBO9780511615634

King, J. (2002). Single –member districts and the representation of women in American state legislatures: The effects of electoral system change. *State Politics & Policy Quarterly*, *2*(2), 161–175. doi:10.1177/153244000200200203

Leefeldt, E. (2016, May 10). *Victim of cyberbullying? Insurance might help*. Retrieved July 5, 2017, from http://www.cbsnews.com/news/victim-of-cyberbullying-insurance-might-help/

Lowi, T. (1964). American business, public policy, case studies, and political theory. *World Politics*, *16*(4), 677–715. doi:10.2307/2009452

McNeal, R. S., Kunkle, S. M., & Bryan, L. D. (2016). State-Level Cyberbullying Policy: Variations in Containing a Digital Problem. In G. Crews (Ed.), *Critical Examinations of School Violence and Disturbance in K-12 Education* (pp. 62–82). Hershey, PA: IGI Global; doi:10.4018/978-1-4666-9935-9.ch005

Meier, K. (1994). *The politics of sin: drugs, alcohol and public policy*. New York: Sharpe.

Mooney, C., & Lee, M. (1995). Legislating morality in the American states: The case of pre-Roe abortion regulation reform. *American Journal of Political Science*, *39*(3), 599–627. doi:10.2307/2111646

Mooney, C., & Lee, M. (2000). The influence of values on consensus and contentious morality policy: U.S. death penalty reform, 1956-82. *The Journal of Politics*, *62*(1), 223–239. doi:10.1111/0022-3816.00011

National Conference of State Legislatures. (2005). *Term limits*. Retrieved March 12, 2007, from http://www.ncsl.org/programs/legman/about/states.htm

National Council of State Legislatures. (2015). *Cyberstalking and cyber harassment laws*. Retrieved September 12, 2015, from http://www.ncsl.org/research/telecommunications-and-information-technology/cyberstalking-and-cyberharassment-laws.aspx

National Telecommunications and Information Administration. (2014). *Exploring the digital nation: Embracing the mobile Internet.* Retrieved February 19, 2015, from http://www.ntia.doc.gov/files/ntia/publications/exploring_the_digital_nation_embracing_the_mobile_internet_10162014.pdf

Nicholson-Crotty, S. (2009). The politics of diffusion: Public Policy in the American states. *The Journal of Politics, 71*(1), 192–205. doi:10.1017/S0022381608090129

NoBullying.com. (2017). *The unforgettable Amanda Todd story.* Retrieved June 8, 2017, from https://nobullying.com/amanda-todd-story/

Partin, R. (2001). Campaign intensity and voter information: A look at gubernatorial contests. *American Politics Research, 29*(2), 115–140. doi:10.1177/1532673X01029002001

Ripley, R., & Franklin, G. (1980). *Congress, the bureaucracy, and public policy.* Homewood, IL: The Dorsey Press.

Rochefort, D., & Cobb, R. (1994). *The politics of problem definition.* Lawrence, KS: Kansas University Press.

United States Census Bureau. (2012a). *2012 Statistical abstract.* Retrieved December 9, 2013, from http://www.census.gov/compendia/statab/

United States Census Bureau. (2012b). *2010 Census urban and rural classification and urban area criteria.* Retrieved November 7, 2013, from http://www.census.gov/geo/reference/ua/urban-rural-2010.html

United States Department of Education. (2013). *Student report of bullying and cyber-bullying: Results from the 2011 school crime supplement to the National Crime Victimization Survey.* Retrieved March 23, 2014, from http://nces.ed.gov/pubs2013/2013329.pdf

United States Department of Health & Human Services. (2015). *What is cyberbullying?* Retrieved January 5, 2015, from http://www.stopbullying.gov/

Walker, J. (1969). The diffusion of innovation among the American states. *The American Political Science Review, 63*(3), 880–899. doi:10.2307/1954434

Will, G. (1993). *Restoration: Congress, term limits, and the recovery of deliberative democracy.* New York: Free Press.

Chapter 6
Policing Online Aggression: Policy Solutions and Challenges

ABSTRACT

Research argues that to address bullying/cyberbullying it will take the larger school community including teachers, families, health professionals, etc. The same can be said for other forms of aggression. This chapter provides an overview of the literature on what each member of the larger community can do to curtail the spread of online aggression. The chapter concludes by examining the effectiveness of recommendations for individuals for protecting themselves from becoming victims of online aggression as well as strategies for parents to protect their children from becoming victims of cyberbullying. Multivariate statistical methods and survey data from the PEW Research Center for the years 2013 and 2014 were used in this analysis.

INTRODUCTION

Media coverage of cyber harassment/bullying seems to have become common place. Nevertheless, not all examples are ones that the average citizen can relate to. In July of 2016, an updated version of the movie Ghostbusters was premiering with all female leads. While the female cast was subjected to misogynist attacks, the harassment was particularly hard on actress Leslie Jones who was subjected to online harassment that was both racist and sexist. The cyber harassment became intense enough that Jones quit Twitter (Fisher & McBride, 2016). Jones was not the only celebrity to be overwhelmed by

DOI: 10.4018/978-1-5225-5285-7.ch006

online harassment. Singer Ed Sheeran quit Twitter in July 2017 because of personal attacks (Ehrbar, 2017). These are only a few examples, as a number of stars have quit social network sites (SNS) such as Twitter or Instagram due to online bullying. Some celebrities take a different tactic to cyber harassment. In 2014, Forrest Rutherford engaged in an online fight with John Popper, the lead singer of Blues Travelers. Rutherford, a social service worker from Kentucky initially had his Twitter account terminated, but since reinstated. Recently, some of Rutherford's old tweets have resurfaced online. In response, Popper has subjected Rutherford to a doxing campaign that includes posting pictures of his home online using Google map along with comments about him and his family. Popper fans have joined in the cyber harassment and Rutherford has reported Popper to Twitter but so far no action has been taken (Cary, 2017).

While not everyone can relate to the cyber-harassment that celebrities like Jones and Sheeran have faced and the online feud with lead singer of Blues Travelers, there are media stories like the "Blue Whale Challenge" that have parents worried. According to the coverage, youth who take part in this challenge are instructed over a 50 day period, to take part in increasingly more dangerous actions. To "win," the participant must commit suicide on the 50th day. Investigation into the claims of the "Blue Whale Challenge" finds there is no evidence of any suicides attributed to this challenge or that it even exists (Patchin, 2017). Even though the challenge appears to be an urban myth, there is still reason for individuals to be concerned. Just as media coverage of celebrities becoming victims of online aggression seem to becoming more common, stories such as the suicides of teens including Megan Meier and Ryan Halligan are not going away. The media has continued to report on similar tragedies. For example, on August 1, 2017, the parents of Mallory Grossman held a news conference to announce that they were suing the Rockaway Township School District in New Jersey for neglect. Mallory, a 6th grader at Copeland Middle School in Rockaway committed suicide at age 12 on June 14, 2017. The lawsuit contends that their daughter had been subjected to bullying (in person) at school, in text messages and through Instagram and Snapchat. Furthermore, the parents stated that while they contacted the school about bullying, administrators failed to follow through with proper district procedures (Stump, 2017). Another recent example is the suicide of 18 year old Conrad Roy III in 2014. Conrad had a history of "threatening to commit suicide" and he did after receiving a series of text messages from girlfriend Michelle Carter pushing him to follow through with his threat. On August 3, 2017, Carter was sentenced to 15 months in jail for involuntary manslaughter but is free pending her appeal (Anderson & Ransom, 2017). Like many bullying/cyberbullying cases, there are legal questions over whether Carter should be prosecuted. As discussed in Chapter 3, laws meant to address bullying are often constrained by the 1st Amendment right to free speech. What actions can be taken to address cyber aggression? Starting with law enforcement, this chapter provides a literature overview on what can be done by each group of stakeholders.

LAW ENFORCEMENT AND LEGISLATURES

One common story told by victims of cybercrime stalking, harassment and nonconsensual pornography (NCP) is frustration with law enforcement. Part of the problem is that the Internet makes it difficult for police to track online criminals. This is, in part, because advances in messaging tools are enabling perpetrators to stay anonymous. For example, a 2016 study by the Pew Research Center found that 24% of smartphone users had messaging apps such as Snapchat or Wickr which automatically deleted sent messages. In addition, 5% of smartphone users have apps such as YikYak or Whisper that allows them to both chat and post messages anonymously (Greenwood, Perrin, & Dugger, 2016). Cybercriminals can also simply create multiple fake email accounts or use a computer at a public library. Even if law enforcement can identify the Internet protocol (IP) address of a computer, it can be difficult to establish who used it. It also takes time and money to investigate cybercrimes. Once investigators identify an IP, they must get a search warrant for a computer or smartphone and send it to criminal lab for examination. Given the resources needed for such an investigation, local law enforcement may limit their efforts to more serious cases (Jany, 2016).

In addition, there is a legal debate over whether suspects in cybercrime can be forced to provide passwords to either unlock or decrypt devices seized using a warrant. There have only been a few court decisions since 2007 concerning the limits of law enforcement in this area and the decisions have been inconsistent. In 2012, the 11th Circuit Court of Appeals ruled that forced decryption constitutes a violation of the defendant's Fifth Amendment rights. Similar, in 2013, a federal judge would not allow law enforcement to force a suspected child pornographer to decrypt his laptop. However, in 2016, a Florida state court ruled that the government could force an individual accused of taking pictures up women skirts to turn over the password to his laptop (Farivar, 2017). Currently, there is not a consensus among judges regarding whether making a defendant provide the passwords to their devises violates the Fifth Amendment. Is there an alternative to forcing an individual to turn over his/her password(s)? The government has, in several high-profile cases, paid a digital forensic firm to get information off of a device. For example, in 2017, the FBI paid $18,000 to Cellebrite, a forensic firm to get information off of an iPhone in a high-profile sexploitation case targeting socialite, Julianna Goddard (Farivar, 2017). While the FBI may pay for assistance in certain high-profile cases, this may be beyond the ability of most local law enforcement agencies.

Victims of cybercrimes often deal with members of law enforcement not familiar with cybercrimes or even the cyber aggression laws in their state. For example, when journalist Catherine Mayer reported that she had received bomb threats, the responding police officers did not understand about Twitter or IP addresses. She was advised to unplug her computer. Katie Beech, a D.J., had a similar response from law enforcement after asking for help when an obsessed fan cyber stalked her. After a judge granted a temporary restraining order, the police sought to serve her alleged stalker; after visiting his residence over a two-day period gave up on serving him. When Beech told police she could pinpoint his exact location using his Twitter updates, she was asked, "What is Twitter" (Beech, 2016)?

How can law enforcement improve its response to cyber aggression? One person trying to answer that question is Congresswomen Katherine Clark who represents the fifth district of Massachusetts. She has been subjected to sexist Tweets since being elected in 2013 as well as being a victim of swatting (Buxton, 2017). As a victim of cyber aggression, Congresswoman Clark has become active in promoting legislation to combat these crimes including sponsoring the Interstate Sextortion Prevention Act, the Interstate Swatting Hoax Act and the Prioritizing Online Threat Enforcement Act of 2015 (Buxton, 2017). Two of the acts help address the problems local law enforcement face when dealing with acts of cyber aggression: limited resources (both time and money) and the lack of training. The first act is the Cybercrime Enforcement Training Assistance Act of 2016 that would provide grants to state and local governments for preventing, enforcing and prosecuting cybercriminals. It would also provide grant money to nonprofit cybercrime organization to conduct research on cybercrimes and provide technical support to law enforcement. The second, the Prioritizing Online Threat Enforcement Act of 2015 would provide money for at least 10 additional agents of the

FBI who would support the Criminal Division of the Department of Justice (DOJ). These agents would help with the investigation and prosecution of cybercrimes against individuals. While both of these bills died in committee, they represent the type of government policy that would help get at the underlying problems associated with investigation and prosecution of cyber aggression, such as cyber harassment and cyber stalking.

Cyber aggression describes a variety of crimes. This discussion on law enforcement concerns crimes such as cyberstalking and harassment primarily committed against adults. Local schools are the agencies most often designated to address cybercrimes between children such as bullying/cyberbullying (see Chapter 3). The literature (Brown, 2015) suggests response by law enforcement can be improved through money for training or even special units within departments that specialize in cybercrimes. The speed at which the Internet changes and the inability of law enforcement to stay ahead of acts of online aggression is made evident by the request for more research into cyber aggression and related crimes is a common theme in the literature. Even though cyberstalking was reported in about ¼ of the cases identified through survey data, it is hypothesized that as technology grows more sophisticated so will the use of technology as a tool, and will be more prevalent. The dark side of criminal justice statistics, essentially unreported crimes, is problematic in understanding the prevalence of stalking and crimes of aggression. Based on responses that many victims have encountered when reporting an instance of cyber harassment/stalking, it may seem like a waste of time. Limited reporting does not allow us to understand the effect these crimes have on victims, especially those from historically minimalized populations such as gays, lesbians, bisexual, and transgendered (see Chapter 2). The issue suggests that the justice system and ancillary agencies specifically dealing with vulnerable populations become more accessible to those at greatest risk. Law enforcement coordinating efforts with ancillary agencies such as domestic violence programs offered by federal grants is one growing approach. Chapter 3 discusses the Services, Training, Officers, Prosecutors Violence against Women Formula Grant Program [STOP], and the Grants to Encourage Arrest Policies and Enforcement of Protection Orders Program (United States Department of Justice, 2015; Congressional Research Service, 2015).

Legislative action is designed to regulate behavior that violates our societal norms, essentially a codification of actions that are deemed too egregious for a civilized society. As such, the criminal justice system becomes involved after the fact of personal harm and sometimes death. One can argue, as Deterrence Theory suggests, that the goal of legislation is to deter future acts of victimization while appropriately and proportionally responding to the sanctioning of the victimization that has already occurred. In that context, legislation that approaches the ideal legislative model defined by the Stalking Resource Center is desirable in all states, territories and jurisdictions. Specialized or therapeutic courts such as domestic violence courts focus on a particular type of offense and a particular type of offender. Furthermore, these types of courts wrap around supervision and treatment services within a due process context. Domestic violence courts emerged in the 1990's with the dual goals of victim assistance and offender accountability. According to a study conducted by Hart and Klein (2013), the outcomes of cases processed through domestic violence courts are conflicting. While some domestic violence courts have seen a reduction in recidivism (relative to re-abuse); others have not (Hart & Klein, 2013, p 575). Nevertheless, we believe that therapeutic jurisprudence courts remain a beneficial and promising alternative to traditional court processing of intimate partner violence cases.

In addition, and consistent with the ideology advocated by the CDC is the goal of prevention. Prevention can be accomplished in many ways. Importantly, it is teaching children to recognize that violence perpetrated by one gender to dominate and control the other is not appropriate or acceptable behavior. Secondly, it is about working within communities and community agencies and organizations to develop strategies promoting respectful and nonviolent relationships. Prevention is about developing a community's capacity to change traditional attitudes about gender stereotypes and beliefs that reinforce the silence and shame that traditionally have hindered the reporting of these types of crimes. Finally, it is about continuing research efforts into understanding the causes of intimate partner violence and identifying strategies that will successfully interrupt the cycle of violence in abusive relationships.

ONLINE ACTIVITIES AND CYBER AGGRESSION

There are many support groups providing advice on how to protect yourself from cyber aggression and what to do if victimized. Some of the suggestions are common sense, including not revealing personal information online such as your home address, contacting web host to report a problem, and not to engage with the cyber attacker (Jany, 2016; Moore, 2015). Some may be less obvious such as saving copies of all forms of communication as evidence in case you seek a restraining order, find the perpetrators IP address, and if you are about to leave a spouse or significant relationship change all passwords to ones that cannot be guessed (Jany, 2016; Moore, 2015). Finally, since non-consenting pornography cases are often still treated as copyright violations, it is important that you copyright your online images such as selfies especially ones that contain explicit images (Hern, 2016). In the next section, the impact of several of these suggestions will be examined using multivariate statistical methods. More specifically, it will examine the impact of sharing personal information and efforts to mask your online presence from those who might engage in acts of online aggression on the likelihood of becoming a victim of cyberaggression.

Empirical Model: Data and Methods

Data

This evaluation of online behavior associated with online aggression utilizes the *2013 Pew Anonymity (omnibus) Survey* conducted for the Pew Internet & American Life Project, by the Princeton Survey Research Associates. The Pew Internet & American Life Project is part of the Pew Research Center, a nonpartisan, nonprofit group that conducts studies and provides information regarding factors that shape American society. Their surveys are random digit dial national telephone surveys limited to individuals 18 years or older and live in the continental United States. Participants of this survey were asked questions regarding anonymity online. This study is based on a survey conducted for the Internet and American Life Project with a sample size (n) of 1,002 adults living in the United States collected from July 11 to July 14, 2013.

Online Activities

This analysis begins by exploring the online activities associated with an individual's likelihood of being a victim of cyber aggression. Independent and control variables were selected based on findings from prior research. Four dependent variables representing different forms of cyber aggression were used. For each, the respondent was asked if his/her online activities had a resulted in one of several negative outcomes. The first outcome was, "Had an email or social networking account of yours been compromised or taken over without your permission by someone else?" The next variable was constructed using the question, "Had your reputation been damaged because of something that happened online. The third dependent variable was constructed using the question, "Had you been stalked or harassed online?" The final dependent variable was constructed using the question, "Had something happened online that led you into physical danger." Each variable is coded 1 if an individual responded yes and 0 otherwise.

Six main independent variables were included. The first two represent actions individuals might take to protect themselves from online aggression where the other four are actions that could increase the likelihood of cyber aggression. The first is a dummy variable based on the question, "Have you ever tried to use the Internet in ways that keep certain friends from being able to see what you have read, watched or posted online?" Similarly, the second independent variable constructed using the question, "Have you ever tried to use the Internet in ways that keep people who might criticize, harass or target you from being able to see what you have read, watched or posted online?" These two activities are expected to decrease the likelihood of becoming a victim of cyber aggression. The remaining four were added based on literature ((Mitchell, Wolak & Finkelhor, 2007; Wolak, Mitchell, & Finkelhor, 2007a) finding that using SNS and making personal information available online is associated with a greater likelihood of victimization. The next two variables were constructed using the questions, "Do you ever use social networking sites like Facebook, LinkedIn or Google Plus?" and "Do you ever use Twitter?" The final two main independent variables were constructed using the questions, "Is a photo of you available on the Internet for others to see?" and "Is information regarding your political party or party affiliation available on the Internet for others to see?" Each variable is coded 1 if an individual responded yes and 0 otherwise.

Control variables include a family income measure based on a 9-point scale where 1 indicated that family income ranges from $0 to $10,000 and 9 signifies a family income of $150,000 or more. Education is measured using a 7-point scale ranging from eighth-grade education or less to post-graduate training/ professional training that would include Masters/ Ph.D., law and medical school. Age is measured in years. Gender is measured using a binary variable coded 1 for male and 0 for female. Younger respondents and females are expected to more likely be victim of cyber aggression because they have found to be more likely to use SNS (Pew Research Center, 2017). To control for race and ethnicity, dummy variables were included for African Americans, Latinos, and others with non-Hispanic Whites as the reference group. Measures of race/ethnicity were included because survey research finds that non-Whites are more likely to be targeted online because of their race. According to a 2017 Pew Research Center survey, 25% of Blacks, 10% of Hispanics and 3% of non-Hispanic Whites report being targeted online because of race/ethnicity. Furthermore, this survey reports that when asked if they thought online harassment was a major problem, more Blacks (74%) and Hispanics (72%) agreed with this statement than non-Hispanic Whites (57%) (Duggan, 2017). Since the National Telecommunication and Information Administration (NTIA) finds regional differences in broadband and Internet access, dummy variables for the Midwest, Northeast and West regions were included to control for regional differences (NTIA, 2014).

Findings and Discussion

In Tables 1 and 2, the dependent variables were coded so that higher scores are associated with increased likelihood of being a victim of cyber aggression. Since the dependent variables are binary, the models are estimated using logistic regression. Regional indicators as well as measures for race/ethnicity and gender were not found to be significantly related to any of the forms of cyber aggression.

There was limited support for the argument that education and income impact the likelihood of being a victim. Those with higher education levels were more likely to have an email or social networking account compromised or taken over without permission by someone else but were less likely to have done something online that led to danger offline. Those with lower income levels were less likely to have an email or social networking account compromised or taken over by someone else and were less likely to have been a victim of cyberstalking or cyber-harassment. This may be because they are less likely to have Internet access. These findings suggest that future surveys should distinguish between different forms of cyber aggression. It is not surprising that younger respondents were more likely to have their reputation damaged online, have been cyberstalked or harassed and done something that led to danger offline. Younger individuals are more likely to have Internet access and be on social media sites (Pew Research Center, 2017).

Although the results for age was not surprising, other findings were not expected. One outcome was not anticipated was the finding that using the Internet in such a way as to hide your activities from certain friends or those that might target or bully you was positively associated with being a victim of cyber aggression. This finding may be because the dataset is cross sectional; it is difficult to establish time order. Individuals may have taken these actions

Table 1. Online activities and cyber aggression, 2013

Variables	SNS or Email Compromised		Reputation Damaged	
	β (se)	p>lzl	β (se)	p>lzl
Environmental Variables				
Midwest	-.380(.328)	.247	-.036(.605)	.953
Northeast	.279(.324)	.388	-.661(.739)	.371
West	-.283(.298)	.341	-.085(.557)	.879
Individual Level Variables				
Age	-.005(.008)	.488	**- .042(.016)**	**.009**
Male	-.128(.232)	.581	.350(.461)	.446
Latino	-.267(.402)	.506	-.183(.702)	.794
Black	-.030(.288)	.939	-.566(.815)	.487
Other	-.311(.430)	.470	-1.235(1.121)	.271
Education	**.166(.072)**	**.020**	-.141(.148)	.342
Income	**-.148(.053)**	**.005**	-.161(.102)	.113
Online Activities				
Uses SNS	**.655(.364)**	**.072**	-.682(.642)	.288
Uses Twitter	.229(.283)	.418	-.758(.635)	.232
Photo Online	**.562(.332)**	**.091**	-.135(.594)	.820
Party Affiliation Online	.235(.260)	.365	**.970(.495)**	**.050**
Hide Activities from Friends	**.681(.287)**	**.018**	**1.052(.532)**	**.048**
Hide Activities from Enemies	**.698(.293)**	**.017**	.641(.549)	.243
Constant	**-2.316(.653)**	**.000**	.132(1.144)	.908
Pseudo R2	.1146		.1668	
LR Chi^2 (16)	63.73	.0000	44.51	.00002
N	550		555	

Sources: *2013 Pew Anonymity (omnibus) Survey.* Logistic regression estimates with standard errors in parentheses. Reported probabilities are based on two-tailed tests. Statistically significant coefficients at .10 or less in bold.

after becoming a victim of cyber aggression. Another unexpected finding was that placing personal information online was only marginally related to cyber aggression. Having a copy of your photo online was associated with having your email or social network account compromised while having your party affiliation online was associated with having your reputation damaged. Future research should examine other forms of personal information such as home address or phone number for their relationship to cyber aggression. These findings may also be time dependent; this survey was taken in 2013. Given the heightened political rhetoric of the 2016 election, it is possible

Table 2. Online activities and cyber aggression, 2013

Variables	Stalked or Harassed		Physical Danger	
	β (se)	p>|z|	β (se)	p>|z|
Environmental Variables				
Midwest	-.443(.492)	.368	.244(.801)	.761
Northeast	.052(.479)	.913	.424(.772)	.583
West	-.599(.457)	.190	-.304(.810)	.707
Individual Level Variables				
Age	**-.030(.012)**	**.015**	**- .072(.028)**	**.010**
Male	.102(.349)	.769	-.440(.602)	.465
Latino	.309(.493)	.531	.148(.768)	.847
Black	-.413(.633)	.514	-1.001(1.185)	.398
Other	-.208(.565)	.713	-1.462(1.173)	.212
Education	- .085(.111)	.441	**-.532(.207)**	**.032**
Income	**-.243(.078)**	**.002**	-.095(.139)	.494
Online Activities				
Uses SNS	.734(.685)	.248	.342(1.195)	.775
Uses Twitter	.229(.283)	.439	.608(.655)	.353
Photo Online	.413(.550)	.453	-.090(.969)	.925
Party Affiliation Online	.525(.380)	.168	-.121(.674)	.857
Hide Activities from Friends	**1.283(.388)**	**.001**	.503(.690)	.466
Hide Activities from Enemies	.547(.400)	.171	**1.389(.677)**	**.040**
Constant	-1.114(1.028)	.278	1.068(1.903)	.574
Pseudo R2	.2420		.3069	
LR Chi² (16)	79.07	.0000	44.51	.00002
N	555		555	

Sources: *2013 Pew Anonymity (omnibus) Survey.* Logistic regression estimates with standard errors in parentheses. Reported probabilities are based on two-tailed tests. Statistically significant coefficients at .10 or less in bold.

that a 2016 survey might find greater association between being a victim of cyber-aggression and having your political affiliation listed online.

PARENTAL INTERVENTION

If your child is a victim of cyberbullying, you may be dealing with the local school district instead of local law enforcement. Most states have passed laws requiring school districts to have anti-bullying policies in place. Even so, research suggests that parents are still the first line of defense in protecting their children. How can parents protect their children from acts of cyber aggression? There is no shortage of advice from online support groups. Research on traditional and online bullying however provides mixed results on the effectiveness of some of the recommendations. A number of studies (Mitchell, Wolak & Finkelhor, 2007; Mesch, 2009; Korchmaros, Mitchell, & Ybarra, 2014) find that the more time spent online including interacting on social network sites increases the likelihood of being a victim of cyber aggression. These findings are consistent with those in models from Chapter 2. This suggests that limiting children's use of the Internet could be a possible strategy for protecting them from cyber aggression. Given the availability of access to the Internet through school, the library, on smartphones or even a friend's house, this could be an increasingly difficult strategy to implement.

Another precaution parents can take is to monitor online behavior. There is mixed results on the effectiveness of this strategy. Dishion, Nelson, and Bullock (2004) argue that monitoring does protect children but not through the mechanism we might think. Instead of protecting children through limiting Internet use; it makes parents aware of potentially bad peer relationships so that they are more likely to intervene. Research (Pepler, Jiange, Craig, & Connolly, 2008; Korchmaros, Mitchell, & Ybarra, 2014) provides support for the argument that monitoring reduces the likelihood that a child will become a victim of cyber aggression. In addition, Lwin, Stanaland and Miyazaki (2008) found parent monitoring reduces the likelihood that a child would engage in risky online activities, but they argue this preventative measure becomes less effective as children age. However, others (Marcum, Ricketts, & Ricketts, 2010; Moore, Guntupalli, & Lee, 2010) did not find any relationship between parental monitoring and likelihood of a youth becoming a victim of cyber aggression. Furthermore, Fleming et al. (2006) found that filters and blocking software did not prevent children from accessing inappropriate material nor did they prevent children from becoming victims of cyberbullying. Finally, many authors (Berson, Berson, & Ferron, 2002; Cassidy, Jackson, & Brown, 2009; Moore, Guntupalli, & Lee, 2010) found that maintaining an open dialogue with one's children is the best approach. They argue that youth are less likely to take part in risky online activities if they have a positive relationship with their parents that include ongoing discussions about Internet use and its risks.

The literature on protecting children from online aggression doesn't provide clear answers to parents. Which measures are more likely to work? The next section uses survey data and multivariate statistical analysis to test which forms of parental intervention are more likely to protect children from online victimization.

Empirical Model: Data and Methods

Data

In examining the importance of parental oversight in preventing cyberbullying, this study utilizes the *2014 Teen Relationship Survey* conducted for the Pew Research Center, by GfK Group (formally known as the Knowledge Network). This is the same dataset used in Chapter 2 to examine the relationship between sexual orientation and cyberbullying. The Pew Research Center is a nonpartisan, nonprofit group that conducts studies and provides information regarding factors that shape American society. Participants of this survey were asked questions regarding online relationships. This study is based on a survey conducted for the Pew Research Center with a sample size (n) of 1,060 teens (ages 13 to 17) and their parents living in the United States collected from September 25, 2014 to October 9, 2014 and from February 10, 2015 to March 16, 2015. Data was collected using a probability-based survey conducted online.

Cyberbullying and Parental Intervention

The models in Table 3, examines whether parental monitoring and intervention lessen the likelihood of being cyberbullied. It explores the same two forms of online bullying examined in Chapter 2: purposely excluding someone from a group and spreading rumors. As in Table 1 of Chapter 2, the first form of cyberbullying is measured using a dummy variable constructed using the following question: "Do you ever experience any of the following on social media: people posting about things you weren't invited to?" The second form of cyberbullying is measured using a dummy variable constructed using the following question: "Do you ever experience any of the following on social media: people posting things about you that you can't change or control?" The variable acts as a measure of rumor spreading. Both variables are coded 1 if an individual responded yes and 0 otherwise. The independent and control variables are the same as those included in Table 1 of Chapter 2. To control for the impact of parental monitoring and intervention actions, five measures were included. The first parental activity was measured using a dummy variable constructed using the following question: "Have you ever done the following: used parental controls or other technological means of blocking, filtering or monitoring your child's online activities?" The second was constructed using the question, "Have you ever done the following: checked which websites your child has visited?" The third measure was based on the question, "Have you ever done the following: checked your child's profile on a social networking site?" The fourth measure used the question, "Have you ever done the following: taken away your child's cell phone or Internet privileges as punishment?" Finally, the last measure of parental control is

Table 3. Parental intervention and cyberbullying, 2014

Variables	Social Exclusion		Rumor Spreading	
	β (se)	p>|z|	β (se)	p>|z|
Parental Variables				
Use Parental Controls	.086(.174)	.623	.227(.174)	.192
Check Websites Visited	-.249(.185)	.178	-.142(.186)	.446
Check Profile on SNS	.132(.187)	.480	.047(.188)	.805
Take Away Access	.136(.178)	.443	**.399(.182)**	**.029**
Limit Use	.039(.176)	.826	.104(.178)	.556
Environmental Variables				
South	.163(.200)	.416	-.074(.202)	.714
Northeast	.373(.245)	.128	.217(.240)	.366
West	.098(.227)	.665	.212(.228)	.354
Individual Level Variables				
Age	.056(.056)	.317	**.145(.056)**	**.010**
Male	**-.316(.151)**	**.036**	-.144(.152)	.344
Latino	**-.452(.207)**	**.029**	**-.686(.215)**	**.001**
Black	-.397(.265)	.135	**-.478(.274)**	**.080**
Other	-.044(.263)	.865	-.121(.261)	.644
Family Income	.023(.017)	.188	.016(.018)	.368
Sexual Orientation	**.640(.281)**	**.023**	**.478(.271)**	**.078**
Internet Usage	**.428(.109)**	**.000**	**.215(.109)**	**.049**
Constant	**-3.103(1.073)**	**.004**	**-4.008(1.092)**	**.000**
Pseudo R2	.0407		.0396	
LR Chi² (16)	43.03	.0003	31.06	.0005
N	767		767	

Sources: *2014 Teen Relationship Survey*. Logistic regression estimates with standard errors in parentheses. Reported probabilities are based on two-tailed tests. Statistically significant coefficients at .10 or less in bold.

constructed using the question, "Have you ever done the following: limited the amount of time or times of day when your child can go online?" Each measure was coded 1 if the parents responded yes and 0 otherwise.

Findings and Discussion

In Table 3, the dependent variables were coded so that higher scores are associated with increased likelihood of being cyberbullied. Since the dependent variables are binary, the models are estimated using logistic regression. The addition of the parental intervention variables had little impact on the significance of the other variables found in Table 1 of Chapter 2. The only findings to change from Table 1 of Chapter 2 and Table 3 was for the variable Black in the model for spreading rumors. In Table 1 of Chapter 2, the variable, Black, was not significant but in Table 3, African American youth were less likely to experience having rumors spread about them online compared to non-Hispanic Whites when parental controls were included.

The findings for race in both Table 1 of Chapter 2 and 3 do not help to clarify why there is inconsistency in the findings on race in the literature. Many studies do not include controls for race and others such as Hinduja and Patchin (2008) included measures of race but did not find a relationship between race and the chance of being a victim of cyberbullying. One study, Zweig, et al (2013) found that Whites were more likely to be the victim of cyberbullying. However, more recent survey research finds that non-Whites are much more likely to be targeted online because of their race/ethnicity (Duggan, 2017). The age of these studies may provide some clue to why the findings are inconsistent among studies. Using the Internet does not in itself increase the likelihood of becoming a victim of cyber aggression but instead, it is the result of taking part in risky activities including sharing personal information with strangers on SNS. Some earlier studies would not have been able to control for social networking; recall when the following technological advances took place, Facebook (2004), Twitter (2006) and smartphones (2007). Although the findings on race are consistent with Zweig, et al (2013) study on teens, they are inconsistent on what we know about cyber aggression in general. Other results were not consistent with previous research. Neither regional indicators nor family income were found to be significantly related to either form of bullying and there was limited support for the argument that gender, and age impact the likelihood of being cyberbullied. In addition, females were more likely to be victims of social exclusion while older teenagers were more likely to be victims of gossiping. These findings suggest that future surveys should distinguish between different forms of cyberbullying. Consistent with research on traditional bullying, certain populations (e.g. females) are more likely to be victims but it is dependent on the type of cyberbullying.

Consistent with the literature and the findings in Chapter 2, both forms of cyberbullying examined in this study (being excluded and having rumors spread about you) were found to be positively associated with the increased Internet use and being a person who identifies as LGBTQ. Furthermore, the dependent variables are based on questions regarding social media use. Prior research (Mitchell, Finkelhor, & Wolak, 2003; Wolak, Mitchell, & Finkelhor, 2007a; Marcum, 2008) argues that it is not simply the amount of time spent online that increases the chance of being bullied but specific online activities such as communicating with others and sharing personal information. Since the dependent variables concern being bullied based on social media use, the results also support this earlier research. The finding that the amount of time spent online is significantly related to being cyberbullied would support the argument that parents should limited online usage. This might be impractical in practice because of the availability of online access at school, in libraries, through smartphones and even at a friend's house. The models in Table 3 provided limited support for the argument that parental monitoring protects children against cyberbullying. The only significant relationship was between having rumors spread about an individual and parents having taken away a child's cell phone or Internet privileges as punishment. The relationship between these variables is positive suggesting that taking away privileges increases that likelihood of being a victim of rumormongering. This result does not make sense, on its face. Because the dataset is cross sectional, it can be difficult to establish time order. A more logical explanation is that parents took away Internet privileges to protect their child after they became a victim of cyberbullying.

What do these findings and the results of other studies suggest parents do to protect their children from cyberbullying? Marcum (2008) recommends that one of the best things parents can do is discuss with their children how to use the Internet safely and which steps to take in order avoid become a victim. Similarly, a number of studies (Hirschi, 2001; Ybarra, & Mitchell, 2004; Wang, Iannotti, Nansal, 2009) find that having a strong bond with parents lessens the chances of becoming a victim. This strong relationship may result in children confiding in parents about their friends and activities. Parents will be more able to steer children away from potentially dangerous circumstances if they have a clear line of communication. Consider the case of Jessica Logan, who committed suicide in 2008 after her former boyfriend shared nude photos that she had sent him. She was harassed constantly at school after her pictures were circulated. She did not tell her parents what was going on and they were not aware there was a problem until they got a note from her school that Jessica had been skipping classes (NoBullying.com, 2015a; Powell, 2016). Had her parents realized sooner that there was a problem, they could have intervened quicker and provided emotional support. Following her death, the state of Ohio passed H.B.116 (The Jessica Logan Act) in 2012; one of the provisions of the law is that parents are notified if their child is being bullied at school.

If parents are aware that their child is a victim of cyber aggression, there are several steps that have been suggested they use to help minimize the harm. These include talking to school administrators, changing contact information (for example create a new email address), blocking tormenters, telling your child not to engage with bullies and contacting the web host to report a problem (NoBullying.com, 2015b). Livingstone et al (2011) found that coping strategies including blocking bullies, disconnecting from the Internet and deleting nasty messages could help minimize some of the pain associated with being a cyber victim. Of these three approaches, blocking was found to be the most effective.

SCHOOL ANTI-BULLYING PROGRAMS

If your child is being cyberbullied, the local government agency you are most likely to work with is the local school district. As discussed in Chapter 4, cyberbullying program are far from consistent. There are three basic elements to addressing cyberbullying and while some school districts may incorporate all of the elements, some may only have one. The first consists of approaches meant to prevent cyberbullying including those teaching students how to use the Internet safely and those focusing on a positive school climate. Hawkins, Pepler, and Craig (2001) find that intervention by bystanders can also quickly deescalate the situation. With that in mind, a number of schools have adopted mentoring programs where student leaders are taught how to step in as a bystander and deescalate a bullying situation. This strategy is being fostered by the #IAmAWitness campaign undertaken by a coalition consisting of the Ad Council, media, corporate and nonprofit groups. The coalition has created a series of emoji, GIFs and stickers that individuals can post when they witness bullying online. There are also emoji and GIFs that can be sent to online victims to signify that they have your support (IWitnessBullying.org, 2017). Another example of this approach is the 'Sit With Us' App invented by high-schooler Natalie Hampton. One example of how students are bullied at school is through being ostracized by their peers and forced to eat alone. This App helps end this form of bullying through linking students to individuals at their school who are open to having others join them for lunch (Wanshel, 2016). The second component consists of procedures that deal with traditional bullying and cyberbullying after it occurs. These actions can include school expulsion and exclusion interventions. They also include therapeutic treatment programs such as anger management classes, programs that focus on raising self-esteem, social-emotional learning, or skill building. The final component of an anti-bullying program is approaches to help the victim manage. These may include peer support programs and teaching coping strategies (Perren, et al., 2012).

The review in Chapter 4 presented several conclusions regarding school anti-bullying programs. The first was that there are no one-size-fits-all approach. Demographical, attitudinal and social differences of a jurisdiction must be factored in, when developing a program. Another finding was that whatever program is adopted, it must have the commitment of stakeholders to see that implementation of the policy is carried out. In a number of high profile cases of teenage suicide, a commonality was that teachers, administrators and other school personal knew about instances of ongoing bullying but failed to do anything about it. A final conclusion is that a program is more likely to succeed when stakeholders besides school personal are included in the process. This group of stakeholders should include parents.

FUTURE RESEARCH DIRECTIONS

Secondary data analysis was used in this chapter to examine the relationship between online bullying and parental intervention in Table 3. Secondary data analysis was also used in Tables 1 and 2 to examine online behavior and the likelihood of becoming a victim of cyber aggression. Like all forms of data collection, this method has advantages and disadvantages. One disadvantage is that the data was collected by someone else for their needs. While this dataset allowed this research to examine the relationship between parental intervention and two forms of cyberbullying in Table 1, future studies should explore other types of online bullying. It should also contain greater control variables for the presence of physical and learning disabilities in the child along with controls for specific online activities. In addition, the study used in Table 3 was limited by age (13-17) and did not include questions regarding disability. The dataset is also a relatively small size (n=1060). The models in Table 1 and 2 have similar problems. The dataset use in these tables could be improved with controls for sexual orientation and the presence of physical and learning disabilities, and is also limited by a relatively small size (n= 1002). As a final issue, each model in the three tables relies on data from a single time period (a cross-sectional design), suggesting outcomes that are only relevant for a specific period of time. In addition, for certain questions it becomes difficult to establish time order. As the Internet continues to change, so does the chance that cyber aggression will evolve and the chance that an individual or specific population will become victims. The results of a cohort study might identify factors that are consistently associated with online bullying or other forms of cyber aggression while a panel study could illustrate the chance of being a victim at different stages of an individual's life.

CONCLUSION

Today, we do a great deal online including emailing, banking, getting the news, shopping, gaming, checking the weather or even contacting our elected officials. At the same time, being online increases the chance that an individual will be a victim of crimes ranging from phishing to cyberbullying. The media provides many examples of online crimes, giving us reason for concern: the hacking of the DNC email, disruption of the power grid in the Ukraine, the compromising of Target customer data and the exposure of identities of AshleyMadison.com customers. The media has also reported on a number of tragic teen suicides including Megan Meier and Ryan Halligan. This development raises some important questions including, "how should individuals best protect themselves from becoming victims of cybercrimes including forms of cyber aggression?" and "how should local law enforcement agencies and school districts be responding to cyber aggression?" The literatures shows that one key to these problems is applying common sense. Some individuals use easy to remember words like "password" as their password for their email and social networking accounts. It is also not unusual to find individuals making online purchases while connected to public Wi-Fi in places such as the library, airport or coffee shop. Just as we need to avoid making these simple mistakes, we also need to make a conscience effort to not make errors that increase the chance of becoming a victim of cyber aggression. This includes not sharing personal information on social media sites including our home and email addresses. In addition, parents are one of the most important deterrence of their children becoming victims of cyberbullying. One of the best actions parents can take is discussing online safety with their children. Just as you tell your children to "not talk to strangers" you tell them to "not share your personal information with strangers online."

Individuals cannot address online aggression alone. Law enforcement and local school systems also play an important role. Law enforcement has struggled in this area because of lack of training and resources. State legislatures and/or Congress could help mitigate these issues by passing bills similar to Cybercrime Enforcement Training Assistance Act of 2016. School districts vary significantly in their response to cyberbullying/bullying in part because of state law. There is very little consistency from state to state. In 2015, Montana became the last state to adopt a bullying policy and it was strictly symbolic. The law provided a definition of bullying but did not mandate any actions to be taken. On the other hand, states (e.g. Florida, Connecticut and Arkansas) have school anti-bullying laws that include cyberbullying and online harassment, and impose both criminal and school sanctions. Even when states have comprehensive anti-bullying policies, they typically give schools considerable leeway in how they are implemented. What type of program should a local school district put into place? There is no "one-size-fits all" anti-bullying program; school districts should focus on developing a program that works best in their community. That said, there is several research findings that school administrators need keep in mind. The first is that the adopted program must have the commitment of all stakeholders to see that implementation is followed. In a number of high profile cases of teenage suicide, a commonality was that teachers, administrators and other school personal knew about instances of ongoing bullying but failed to respond. In addition, an anti-bullying policy is more likely to succeed if other stakeholders besides school personal are included in the process; included in this group are parents.

REFERENCES

Agatston, P., Kowalski, R., & Limber, S. (2007). Students' perspectives on cyber bullying. *The Journal of Adolescent Health*, *41*(6), S59–S60. doi:10.1016/j.jadohealth.2007.09.003 PMID:18047946

Anderson, T., & Ransom, J. (2017, August 3). *Michelle Carter receives 15 months in jail; will remain free pending appeal.* Retrieved August 3, 2017, from http://www.msn.com/en-us/news/crime/michelle-carter-receives-15-months-in-jail-will-remain-free-pending-appeal/ar-AApls87?li=BBmkt5R&ocid=spartandhp

Beech, K. (2016, June 14). *I had a cyberstalker, and the police wouldn't help me.* Retrieved May 25, 2017, from https://www.nylon.com/articles/cyberstalker-police

Berson, I., Berson, M., & Ferron, J. (2002). Emerging risk of violence in the digital age: Lessons for educators from an online study of adolescent girls in the United States. *Journal of School Violence*, *1*(2), 15–33. doi:10.1300/J202v01n02_04

Brown, C. S. D. (2015). Investigating and prosecuting cyber crime: Forensic dependencies and barriers to justice. *International Journal of Cyber Criminology*, *9*(1), 55–119.

Buxton, M. (2017, March 28). *Meet the congresswomen on the Frontlines of the fight against online abuse.* Retrieved July 19, 2017, from http://www.refinery29.com/2017/03/147268/katherine-clark-online-abuse-interview

Cary, G. (2017, July 31). *John Popper of Blue's Travelers in strange online feud.* Retrieved July 31, 2017, from http://www.newser.com/story/246526/man-claims-blues-traveler-singer-is-trolling-him.html

Cassidy, W., Jackson, M., & Brown, K. (2009). Sticks and stones can break my bones, but how can pixels hurt me? Students' experiences with cyber-bullying. *School Psychology International*, *30*(4), 383–402. doi:10.1177/0143034309106948

Congressional Research Service. (2015). *The violence against women act: Overview, legislation, and federal funding.* Retrieved June 23, 2017, from https://www.crs.gov

Dishion, T., Nelson, S., & Bullock, B. (2004). Premature adolescent autonomy: Parent disengagement and deviant peer process in the amplification of problem behavior. *Journal of Adolescence*, *27*(5), 515–530. doi:10.1016/j.adolescence.2004.06.005 PMID:15475044

Duggan, M. (2017, July 25). *1 in 4 black Americans have faced online harassment because of their race or ethnicity*. Retrieved July 28, 2017, from http://www.pewresearch.org/fact-tank/2017/07/25/1-in-4-black-americans-have-faced-online-harassment-because-of-their-race-or-ethnicity/

Ehrbar, N. (2017, July 4). *Ed Sheeran quit Twitter because people are mean*. Retrieved July 12, 2017, from http://www.cbsnews.com/news/ed-sheeran-quit-twitter-people-are-mean/

Farivar, C. (2017, July 29). *When sextortion suspect refused to unlock her iPhone, the FBI stepped in*. Retrieved July 30, 2017, from https://arstechnica.com/tech-policy/2017/07/with-fbi-and-cellebrites-help-miami-police-bust-into-seized-iphone-6/

Fisher, L., & McBride, B. (2016, July 20). *'Ghostbusters' star Leslie Jones quits Twitter after online harassment*. Retrieved July 20, 2017, from http://abcnews.go.com/Entertainment/ghostbusters-star-leslie-jones-quits-twitter-online-harassment/story?id=40698459

Fleming, M., Greenstreet, S., Cocotti-Muller, K., Elias, D., & Morrison, S. (2006). Safety in cyberspace: Adolescents' safety and exposure online. *Youth & Society*, *38*(2), 135–154. doi:10.1177/0044118X06287858

Greenwood, S., Perrin, A., & Duggan, M. (2016, November 11). *Social media update 2016*. Retrieved December 19, 2016, from http://www.pewinternet.org/2016/11/11/social-media-update-2016/

Hart, B., & Klein, A. (2013). *Practical Implications of Current Intimate Partner Violence Research for Victim Advocates and Service Providers, document no. 244348*. Washington, DC: U.S. Department of Justice.

Hawkins, D., Pepler, D., & Craig, W. (2001). Peer interventions in playground bullying. *Social Development*, *10*, 512–527. doi:10.1111/1467-9507.00178

Hern, A. (2016, May 9). *How revenge porn sites rely on legal loopholes and anonymity*. Retrieved May 15, 2017, from https://www.theguardian.com/technology/2016/may/09/revenge-porn-websites-legal-loopholes-anonymity

Hinduja, S., & Patchin, J. (2008). Cyberbullying: An exploratory analysis of factors related to offending and victimization. *Deviant Behavior*, *29*(2), 129–156. doi:10.1080/01639620701457816

Hirschi, T. (2001). *Causes of delinquency*. Piscataway, NJ: Transaction Publishers.

House of Representatives 2602. (2015-2016). *Prioritizing Online Threat Enforcement Act of 2015.*

House of Representatives 4740. (2015-2016). *Cybercrime Enforcement Training Assistance Act of 2016.*

IWitnessBullying.org. (2017). *Press kit.* Retrieved August 6, 2017, from http://iwitnessbullying.adcouncilkit.org/

Jany, L. (2016, June 8). *In electronic era, stalkers, abusers can be hard to track.* Retrieved May 12, 2017, from http://www.startribune.com/in-electronic-era-stalkers-abusers-can-be-hard-to-track/382278371/

Korchmaros, J., Mitchell, K., & Ybarra, M. (2014). Technology-based interpersonal victimization: Predictors of patterns of victimization over time. *Journal of Interpersonal Violence, 29*(7), 1297–1317. doi:10.1177/0886260513506277 PMID:24269990

Limber, S. (2003). Efforts to address bullying in U.S. schools. *American Journal of Health Education, 34*(5), S-23–S-29. doi:10.1080/19325037.2003.10603589

Livingston, S., Haddon, L., Görzig, A., & Olafsson, K. (2011). Risks and safety on the Internet: The perspective of European children: Full findings and policy implications from the EU Kids Online Survey of 9-16 Year Olds and Their Parents in 25 Countries. London: LSE.

Lwin, M., Stanaland, A., & Miyazaki, A. (2008). Protecting children's privacy online: How parental mediation strategies affect website safeguard effectiveness. *Journal of Retailing, 84*(2), 205–217. doi:10.1016/j.jretai.2008.04.004

Marcum, C. (2008). Identifying potential factors of adolescent online victimization for high school seniors. *International Journal of Cyber Criminology, 2*(2), 346–367.

Marcum, C., Ricketts, M., & Ricketts, M. (2010). Potential factors of online victimization of youth: An online examination of adolescent online behaviors utilizing Routine Activities Theory. *Deviant Behavior, 31*(5), 1–31. doi:10.1080/01639620903004903

Mesch, G. (2009). Parental mediation, online activities, and cyberbullying. *Cyberpsychology & Behavior, 12*(4), 387–393. doi:10.1089/cpb.2009.0068 PMID:19630583

Mitchell, K., Finkelhor, D., & Wolak, J. (2003). The exposure of youth to unwanted sexual material on the internet: A national survey of risk, impact and prevention. *Youth & Society*, *34*(3), 3300–3358. doi:10.1177/0044118X02250123

Mitchell, K., Wolak, J., & Finkelhor, D. (2007). Trends in youth reports of sexual solicitations, harassment and unwanted exposure to pornography on the internet. *The Journal of Adolescent Health*, *40*(2), 116–126. doi:10.1016/j.jadohealth.2006.05.021 PMID:17259051

Moore, A. (2015, October 22). *12 tips to protect yourself from cyberstalking*. Retrieved April, 19, 2017, from https://www.thoughtco.com/tips-to-protect-yourself-from-cyberstalking-3534318

Moore, R., Guntupalli, N., & Lee, T. (2010). Parental regulation and online activities: Examining factors that influence a youth's potential to become a victim of online harassment. *International Journal of Cyber Criminology*, *4*(1&2), 685–698.

National Telecommunication and Information Administration. (2014). *Exploring the digital nation: Embracing the mobile Internet*. Retrieved March 17, 2015, from https://www.ntia.doc.gov/files/ntia/publications/exploring_the_digital_nation_embracing_the_mobile_internet_10162014.pdf

NoBullying.com. (2015a). *Jessica Logan-The rest of the story*. Retrieved July 12, 2017, from https://nobullying.com/jessica-logan/

NoBullying.com. (2015b). *How to protect yourself from cyber bullying today*. Retrieved July 3, 2017, from https://nobullying.com/how-to-protect-yourself-from-cyber-bullying/

Patchin, J. (2017, May 16). *Blue Whale Challenge*. Retrieved June 11, 2017, from https://cyberbullying.org/blue-whale-challenge

Pepler, D., Jiange, D., Craig, W., & Connolly, J. (2008). Developmental trajectories of bullying and associated factors. *Child Development*, *79*(2), 325–338. doi:10.1111/j.1467-8624.2007.01128.x PMID:18366426

Perren, J., Corcoran, L., Cowie, H., Dehue, F., Garcia, D., McGuckin, C., ... Völlink, T. (2012). Tackling cyberbullying: Review of empirical evidence regarding successful responses by student, parents, and schools. *International Journal of Conflict and Violence*, *6*(2), 283–293.

Pew Research Center. (2017). *Online Harassment 2017*. Retrieved July 20, 2017, from http://www.pewinternet.org/2017/07/11/online-harassment-2017/

Powell, A. (2016). The sexting "epidemic": What can teachers do to prevent disruptions in school caused by sexting? *Journal of Law & Education, 45*(4), 605–612.

Rosen, L., Cheever, N., & Carrier, L. (2008). The association of parenting style and child age and parental limit setting and adolescent Myspace behavior. *Journal of Applied Developmental Psychology, 29*(6), 459–471. doi:10.1016/j.appdev.2008.07.005

Stump, S. (2017, August 2). *Parents of bullied 12-year-old girl who committed suicide: 'We want to honor her.'* Retrieved August 2, 2017, from http://www.msn.com/en-us/health/healthtrending/parents-of-bullied-12-year-old-girl-who-committed-suicide-we-want-to-honor-her/arAApiPSl?li=BBnba9O&ocid=mailsignout

U.S. Department of Justice Office on Violence against Women. (2015). *STOP program 2014 report part A*. Retrieved June, 15, 2017, from http://www.ovw.usdoj.gov

Wang, J., Iannotti, R., & Nansel, T. (2009). School bullying among adolescents in the United States: Physical, verbal, relational, cyber. *The Journal of Adolescent Health, 45*(4), 368–375. doi:10.1016/j.jadohealth.2009.03.021 PMID:19766941

Wanshel, E. (2016). *Teen makes 'Sit With Us' App that helps students find lunch buddies*. Retrieved December 18, 2017, from https://www.huffingtonpost.com/entry/teen-creates-app-sit-with-us-open-welcoming-tables-lunch-bullying_us_57c5802ee4b09cd22d926463

Wolak, J., Mitchell, K., & Finkelhor, D. (2007a). Unwanted and wanted exposure to online pornography in a national sample of youth internet users. *Pediatrics, 119*(2), 247–257. doi:10.1542/peds.2006-1891 PMID:17272613

Worthen, M. (2007). Education policy implications from the expert panel on electronic media and youth violence. *The Journal of Adolescent Health, 41*(6), S61–S63. doi:10.1016/j.jadohealth.2007.09.009 PMID:18047948

Ybarra, M., & Mitchell, K. (2004). Youth engagement in online harassment: Associations with caregiver-child relationships, Internet use, and personal characteristics. *Journal of Adolescence*, *27*(3), 319–336. doi:10.1016/j.adolescence.2004.03.007 PMID:15159091

Zweig, J. M., Dank, M., Lachman, P., & Yahne, J. (2013). *Technology, teen dating violence and abuse, and Bullying*. Urban Institute's Justice Policy Center, Document No. 243296. Retrieved January 9, 2015, from https://www.ncjrs.gov/pdffiles1/nij/grants/243296.pdf

Related Readings

To continue IGI Global's long-standing tradition of advancing innovation through emerging research, please find below a compiled list of recommended IGI Global book chapters and journal articles in the areas of cyber bullying, policy reform, and aggressive behavior. These related readings will provide additional information and guidance to further enrich your knowledge and assist you with your own research.

Abuljadail, M. H., Ha, L., Wang, F., & Yang, L. (2015). What Motivates Online Shoppers to "Like" Brands' Facebook Fan Pages? In A. Mesquita & C. Tsai (Eds.), *Human Behavior, Psychology, and Social Interaction in the Digital Era* (pp. 279–293). Hershey, PA: IGI Global. doi:10.4018/978-1-4666-8450-8.ch014

Adomi, E. E., Eriki, J. A., Tiemo, P. A., & Akpojotor, L. O. (2016). Incidents of Cyberbullying Among Library and Information Science (LIS) Students at Delta State University, Abraka, Nigeria. *International Journal of Digital Literacy and Digital Competence*, *7*(4), 52–63. doi:10.4018/IJDLDC.2016100104

Ahad, A. D., & Anshari, M. (2017). Smartphone Habits Among Youth: Uses and Gratification Theory. *International Journal of Cyber Behavior, Psychology and Learning*, *7*(1), 65–75. doi:10.4018/IJCBPL.2017010105

Akmam, J., & Huq, N. (2016). Living Parallel-ly in Real and Virtual: Internet as an Extension of Self. In A. Novak & I. El-Burki (Eds.), *Defining Identity and the Changing Scope of Culture in the Digital Age* (pp. 230–239). Hershey, PA: IGI Global. doi:10.4018/978-1-5225-0212-8.ch014

Alim, S. (2015). Analysis of Tweets Related to Cyberbullying: Exploring Information Diffusion and Advice Available for Cyberbullying Victims. *International Journal of Cyber Behavior, Psychology and Learning*, *5*(4), 31–52. doi:10.4018/IJCBPL.2015100103

Alim, S. (2016). Cyberbullying in the World of Teenagers and Social Media: A Literature Review. *International Journal of Cyber Behavior, Psychology and Learning*, *6*(2), 68–95. doi:10.4018/IJCBPL.2016040105

Alim, S. (2017). Twitter Profiles of Organisations Fighting Against Cyberbullying and Bullying: An Exploration of Tweet Content, Influence and Reachability. *International Journal of Cyber Behavior, Psychology and Learning*, *7*(3), 37–56. doi:10.4018/IJCBPL.2017070104

Alonso de Escamilla, A. (2015). Answering the New Realities of Stalking. In M. Cruz-Cunha & I. Portela (Eds.), *Handbook of Research on Digital Crime, Cyberspace Security, and Information Assurance* (pp. 67–76). Hershey, PA: IGI Global. doi:10.4018/978-1-4666-6324-4.ch005

Andriakaina, E. (2016). Public History and National Identity: The 1821 Revolution as Metaphor for the "Greek Crisis". In A. Novak & I. El-Burki (Eds.), *Defining Identity and the Changing Scope of Culture in the Digital Age* (pp. 56–79). Hershey, PA: IGI Global. doi:10.4018/978-1-5225-0212-8.ch005

Antonietti, A., Caravita, S. C., Colombo, B., & Simonelli, L. (2015). Blogs' Potentialities in Learning: What Are the Key Variables to Promote Cognitive Empowerment. In A. Mesquita & C. Tsai (Eds.), *Human Behavior, Psychology, and Social Interaction in the Digital Era* (pp. 21–44). Hershey, PA: IGI Global. doi:10.4018/978-1-4666-8450-8.ch002

Arslan, G. (2018). Psychological Maltreatment and Internet Addiction: Is Psychological Maltreatment a Risk Factor? In B. Bozoglan (Ed.), *Psychological, Social, and Cultural Aspects of Internet Addiction* (pp. 90–108). Hershey, PA: IGI Global. doi:10.4018/978-1-5225-3477-8.ch005

Ayscue, L. M. (2016). Perception of Communication in Virtual Learning Environments: What's in It for Them? In B. Baggio (Ed.), *Analyzing Digital Discourse and Human Behavior in Modern Virtual Environments* (pp. 25–39). Hershey, PA: IGI Global. doi:10.4018/978-1-4666-9899-4.ch002

Baggio, B. G. (2016). Why We Would Rather Text than Talk: Personality, Identity, and Anonymity in Modern Virtual Environments. In B. Baggio (Ed.), *Analyzing Digital Discourse and Human Behavior in Modern Virtual Environments* (pp. 110–125). Hershey, PA: IGI Global. doi:10.4018/978-1-4666-9899-4.ch006

Bagwell, T. C., & Jackson, S. L. (2016). The Mode of Information – Due Process of Law and Student Loans: Bills of Attainder Enter the Digital Age. In R. Cropf & T. Bagwell (Eds.), *Ethical Issues and Citizen Rights in the Era of Digital Government Surveillance* (pp. 16–34). Hershey, PA: IGI Global. doi:10.4018/978-1-4666-9905-2.ch002

Bertolotti, T., & Magnani, L. (2015). Cyber-Bullies as Cyborg-Bullies. *International Journal of Technoethics*, *6*(1), 35–44. doi:10.4018/ijt.2015010103

Betts, L. R. (2015). Cyber Bullying Behaviours. In M. Khosrow-Pour (Ed.), *Encyclopedia of Information Science and Technology* (3rd ed.; pp. 6727–6735). Hershey, PA: IGI Global. doi:10.4018/978-1-4666-5888-2.ch661

Betts, L. R. (2018). The Nature of Cyber Bullying Behaviours. In M. Khosrow-Pour, D.B.A. (Ed.), Encyclopedia of Information Science and Technology, Fourth Edition (pp. 4245-4254). Hershey, PA: IGI Global. doi:10.4018/978-1-5225-2255-3.ch368

Betts, L. R., & Spenser, K. A. (2015). "A Large Can of Worms": Teachers' Perceptions of Young People's Technology Use. *International Journal of Cyber Behavior, Psychology and Learning*, *5*(2), 15–29. doi:10.4018/ijcbpl.2015040102

Biradar, R. C., & Nayaka, R. J. (2015). Modern Crypto Systems in Next Generation Networks: Issues and Challenges. In M. Cruz-Cunha & I. Portela (Eds.), *Handbook of Research on Digital Crime, Cyberspace Security, and Information Assurance* (pp. 263–276). Hershey, PA: IGI Global. doi:10.4018/978-1-4666-6324-4.ch017

Bisen, S. S., & Deshpande, Y. (2018). The Impact of the Internet in Twenty-First Century Addictions: An Overview. In B. Bozoglan (Ed.), *Psychological, Social, and Cultural Aspects of Internet Addiction* (pp. 1–19). Hershey, PA: IGI Global. doi:10.4018/978-1-5225-3477-8.ch001

Bishop, J. (2015). Sticks and Stones Will Break My Euros: The Role of EU Law in Dealing with Cyber-Bullying through Sysop-Prerogative. In M. Cruz-Cunha & I. Portela (Eds.), *Handbook of Research on Digital Crime, Cyberspace Security, and Information Assurance* (pp. 424–435). Hershey, PA: IGI Global. doi:10.4018/978-1-4666-6324-4.ch027

Bishop, J. (2015). Trolling Is Not Just a Art. It Is an Science: The Role of Automated Affective Content Screening in Regulating Digital Media and Reducing Risk of Trauma. In M. Cruz-Cunha & I. Portela (Eds.), *Handbook of Research on Digital Crime, Cyberspace Security, and Information Assurance* (pp. 436–450). Hershey, PA: IGI Global. doi:10.4018/978-1-4666-6324-4.ch028

Bishop, J. (2018). Evaluating the Risk of Digital Addiction in Blended Learning Environments: Considering ICT Intensity, Learning Style, and Architecture. In B. Bozoglan (Ed.), *Psychological, Social, and Cultural Aspects of Internet Addiction* (pp. 169–185). Hershey, PA: IGI Global. doi:10.4018/978-1-5225-3477-8.ch009

Boursier, V., & Manna, V. (2018). Problematic Linkages in Adolescents: Italian Adaptation of a Measure for Internet-Related Problems. In B. Bozoglan (Ed.), *Psychological, Social, and Cultural Aspects of Internet Addiction* (pp. 253–282). Hershey, PA: IGI Global. doi:10.4018/978-1-5225-3477-8.ch014

Boylu, A. A., & Günay, G. (2018). Loneliness and Internet Addiction Among University Students. In B. Bozoglan (Ed.), *Psychological, Social, and Cultural Aspects of Internet Addiction* (pp. 109–125). Hershey, PA: IGI Global. doi:10.4018/978-1-5225-3477-8.ch006

Bozoglan, B. (2018). The Role of Family Factors in Internet Addiction Among Children and Adolescents: An Overview. In B. Bozoglan (Ed.), *Psychological, Social, and Cultural Aspects of Internet Addiction* (pp. 146–168). Hershey, PA: IGI Global. doi:10.4018/978-1-5225-3477-8.ch008

Brkljačić, T., Majetić, F., & Wertag, A. (2018). I'm Always Online: Well-Being and Main Sources of Life Dis/Satisfaction of Heavy Internet Users. In B. Bozoglan (Ed.), *Psychological, Social, and Cultural Aspects of Internet Addiction* (pp. 72–89). Hershey, PA: IGI Global. doi:10.4018/978-1-5225-3477-8.ch004

Cacho-Elizondo, S., Shahidi, N., & Tossan, V. (2015). Giving Up Smoking Using SMS Messages on your Mobile Phone. In A. Mesquita & C. Tsai (Eds.), *Human Behavior, Psychology, and Social Interaction in the Digital Era* (pp. 72–94). Hershey, PA: IGI Global. doi:10.4018/978-1-4666-8450-8.ch004

Campbell, M. A., Whiteford, C., Duncanson, K., Spears, B., Butler, D., & Slee, P. T. (2017). Cyberbullying Bystanders: Gender, Grade, and Actions among Primary and Secondary School Students in Australia. *International Journal of Technoethics*, *8*(1), 44–55. doi:10.4018/IJT.2017010104

Carneiro, A. (2015). Defending Information Networks in Cyberspace: Some Notes on Security Needs. In M. Cruz-Cunha & I. Portela (Eds.), *Handbook of Research on Digital Crime, Cyberspace Security, and Information Assurance* (pp. 314–333). Hershey, PA: IGI Global. doi:10.4018/978-1-4666-6324-4.ch020

Casaregola, V. (2016). Who "Screens" Security?: Cultures of Surveillance in Film. In R. Cropf & T. Bagwell (Eds.), *Ethical Issues and Citizen Rights in the Era of Digital Government Surveillance* (pp. 57–76). Hershey, PA: IGI Global. doi:10.4018/978-1-4666-9905-2.ch004

Chabot, Y., Bertaux, A., Kechadi, T., & Nicolle, C. (2015). Event Reconstruction: A State of the Art. In M. Cruz-Cunha & I. Portela (Eds.), *Handbook of Research on Digital Crime, Cyberspace Security, and Information Assurance* (pp. 231–245). Hershey, PA: IGI Global. doi:10.4018/978-1-4666-6324-4.ch015

Chamakiotis, P., & Panteli, N. (2016). The World is your Office: Being Creative in a Global Virtual Organization. In B. Baggio (Ed.), *Analyzing Digital Discourse and Human Behavior in Modern Virtual Environments* (pp. 87–108). Hershey, PA: IGI Global. doi:10.4018/978-1-4666-9899-4.ch005

Chen, B., & Luppicini, R. (2017). The New Era of Bullying: A Phenomenological Study of University Students' Past Experience with Cyberbullying. *International Journal of Cyber Behavior, Psychology and Learning*, *7*(2), 72–90. doi:10.4018/IJCBPL.2017040106

Chen, I. L., & Shen, L. (2016). The Cyberethics, Cybersafety, and Cybersecurity at Schools. *International Journal of Cyber Ethics in Education*, *4*(1), 1–15. doi:10.4018/IJCEE.2016010101

Chen, R., Lin, T., & Xie, T. (2015). Towards Intelligent Window Layout Management: The Role of Mental Map. In A. Mesquita & C. Tsai (Eds.), *Human Behavior, Psychology, and Social Interaction in the Digital Era* (pp. 146–161). Hershey, PA: IGI Global. doi:10.4018/978-1-4666-8450-8.ch007

Ciufudean, C. (2015). Cyber Security Model of Artificial Social System Man-Machine. In M. Cruz-Cunha & I. Portela (Eds.), *Handbook of Research on Digital Crime, Cyberspace Security, and Information Assurance* (pp. 500–514). Hershey, PA: IGI Global. doi:10.4018/978-1-4666-6324-4.ch031

Cline, B. J. (2016). The Electric Soul: Faith, Spirituality, and Ontology in a Digital Age. In A. Novak & I. El-Burki (Eds.), *Defining Identity and the Changing Scope of Culture in the Digital Age* (pp. 251–270). Hershey, PA: IGI Global. doi:10.4018/978-1-5225-0212-8.ch016

Comas-Forgas, R., Sureda-Negre, J., & Calvo-Sastre, A. (2017). Characteristics of Cyberbullying Among Native and Immigrant Secondary Education Students. *International Journal of Cyber Behavior, Psychology and Learning*, *7*(1), 1–17. doi:10.4018/IJCBPL.2017010101

Cooper, V. N., Shahriar, H., & Haddad, H. M. (2015). Development and Mitigation of Android Malware. In M. Cruz-Cunha & I. Portela (Eds.), *Handbook of Research on Digital Crime, Cyberspace Security, and Information Assurance* (pp. 51–66). Hershey, PA: IGI Global. doi:10.4018/978-1-4666-6324-4.ch004

Costa, C., Sousa, C., Rogado, J., & Henriques, S. (2017). Playing Digital Security: Youth Voices on their Digital Rights. *International Journal of Game-Based Learning*, *7*(3), 11–25. doi:10.4018/IJGBL.2017070102

Curado, H. (2015). DNA Databases for Criminal Investigation. In M. Cruz-Cunha & I. Portela (Eds.), *Handbook of Research on Digital Crime, Cyberspace Security, and Information Assurance* (pp. 99–115). Hershey, PA: IGI Global. doi:10.4018/978-1-4666-6324-4.ch007

Dias de Castro, T. S., & Osório, A. J. (2015). Online Violence: Listening to Children's Online Experiences. In M. Cruz-Cunha & I. Portela (Eds.), *Handbook of Research on Digital Crime, Cyberspace Security, and Information Assurance* (pp. 35–50). Hershey, PA: IGI Global. doi:10.4018/978-1-4666-6324-4.ch003

Doherty, C., Lang, M., Deane, J., & Connor, R. (2015). Information Disclosure on Social Networking Sites: An Exploratory Survey of Factors Impacting User Behaviour on Facebook. In M. Cruz-Cunha & I. Portela (Eds.), *Handbook of Research on Digital Crime, Cyberspace Security, and Information Assurance* (pp. 515–532). Hershey, PA: IGI Global. doi:10.4018/978-1-4666-6324-4.ch032

Eckert, S., & Steiner, L. (2016). Feminist Uses of Social Media: Facebook, Twitter, Tumblr, Pinterest, and Instagram. In A. Novak & I. El-Burki (Eds.), *Defining Identity and the Changing Scope of Culture in the Digital Age* (pp. 210–229). Hershey, PA: IGI Global. doi:10.4018/978-1-5225-0212-8.ch013

Edwards, S. B. III. (2016). The Right to Privacy Is Dying: Technology Is Killing It and We Are Letting It Happen. In R. Cropf & T. Bagwell (Eds.), *Ethical Issues and Citizen Rights in the Era of Digital Government Surveillance* (pp. 103–126). Hershey, PA: IGI Global. doi:10.4018/978-1-4666-9905-2.ch006

El-Burki, I. J., & Reynolds, R. R. (2016). It's No Secret Justin Wants to Be Black: Comedy Central's Justin Bieber Roast and Neoliberalism. In A. Novak & I. El-Burki (Eds.), *Defining Identity and the Changing Scope of Culture in the Digital Age* (pp. 15–28). Hershey, PA: IGI Global. doi:10.4018/978-1-5225-0212-8.ch002

Endong, F. P. (2018). Selfie-Objectification as a Facet of the Social Media Craze Among Youths in Nigeria: A Socio-Cultural Discourse. In B. Bozoglan (Ed.), *Psychological, Social, and Cultural Aspects of Internet Addiction* (pp. 236–252). Hershey, PA: IGI Global. doi:10.4018/978-1-5225-3477-8.ch013

Englander, E. K., & McCoy, M. (2017). Pressured Sexting and Revenge Porn in a Sample of Massachusetts Adolescents. *International Journal of Technoethics*, *8*(2), 16–25. doi:10.4018/IJT.2017070102

Fairfax, T., Laing, C., & Vickers, P. (2015). Network Situational Awareness: Sonification and Visualization in the Cyber Battlespace. In M. Cruz-Cunha & I. Portela (Eds.), *Handbook of Research on Digital Crime, Cyberspace Security, and Information Assurance* (pp. 334–349). Hershey, PA: IGI Global. doi:10.4018/978-1-4666-6324-4.ch021

Fang, L., & Ha, L. (2015). Do College Students Benefit from Their Social Media Experience?: Social Media Involvement and Its Impact on College Students' Self-Efficacy Perception. In A. Mesquita & C. Tsai (Eds.), *Human Behavior, Psychology, and Social Interaction in the Digital Era* (pp. 259–278). Hershey, PA: IGI Global. doi:10.4018/978-1-4666-8450-8.ch013

Farmer, L. S. (2016). Using Virtual Environments to Transform Collective Intelligence. In B. Baggio (Ed.), *Analyzing Digital Discourse and Human Behavior in Modern Virtual Environments* (pp. 149–163). Hershey, PA: IGI Global. doi:10.4018/978-1-4666-9899-4.ch008

Firdhous, M. F. (2016). Cyber Espionage: How Safe Are We? In R. Cropf & T. Bagwell (Eds.), *Ethical Issues and Citizen Rights in the Era of Digital Government Surveillance* (pp. 176–207). Hershey, PA: IGI Global. doi:10.4018/978-1-4666-9905-2.ch010

Fiske, R. R. (2016). The Borders of Corruption: Living in the State of Exception. In R. Cropf & T. Bagwell (Eds.), *Ethical Issues and Citizen Rights in the Era of Digital Government Surveillance* (pp. 1–15). Hershey, PA: IGI Global. doi:10.4018/978-1-4666-9905-2.ch001

Force, C. M. (2016). How Middle School Principals of Small Rural Schools Address Cyberbullying. *International Journal of Cyber Behavior, Psychology and Learning*, *6*(1), 27–41. doi:10.4018/IJCBPL.2016010102

Gao, M., Zhao, X., & McJunkin, M. (2016). Adolescents' Experiences of Cyberbullying: Gender, Age and Reasons for Not Reporting to Adults. *International Journal of Cyber Behavior, Psychology and Learning*, *6*(4), 13–27. doi:10.4018/IJCBPL.2016100102

Glasser, D., & Taneja, A. (2015). A Routine Activity Theory-Based Framework for Combating Cybercrime. In M. Cruz-Cunha & I. Portela (Eds.), *Handbook of Research on Digital Crime, Cyberspace Security, and Information Assurance* (pp. 398–406). Hershey, PA: IGI Global. doi:10.4018/978-1-4666-6324-4.ch025

Grzonkowski, S., Vasiliu, L., & Koumpis, A. (2015). Raptor: Early Recognition and Elimination of Network Attacks. In M. Cruz-Cunha & I. Portela (Eds.), *Handbook of Research on Digital Crime, Cyberspace Security, and Information Assurance* (pp. 19–34). Hershey, PA: IGI Global. doi:10.4018/978-1-4666-6324-4.ch002

Gur, S., Blanchard, A. L., & Walker, L. S. (2016). Impacts on Society: Informational and Socio-Emotional Support in Virtual Communities and Online Groups. In B. Baggio (Ed.), *Analyzing Digital Discourse and Human Behavior in Modern Virtual Environments* (pp. 181–195). Hershey, PA: IGI Global. doi:10.4018/978-1-4666-9899-4.ch010

Hacker, K. L., Acquah-Baidoo, B., & Epperson, A. (2016). Reconciling the Needs for National Security and Citizen Privacy in an Age of Surveillance. In R. Cropf & T. Bagwell (Eds.), *Ethical Issues and Citizen Rights in the Era of Digital Government Surveillance* (pp. 78–102). Hershey, PA: IGI Global. doi:10.4018/978-1-4666-9905-2.ch005

Hartzel, K. S., & Gerde, V. W. (2016). Using Duality Theory to Reframe E-Government Challenges. In R. Cropf & T. Bagwell (Eds.), *Ethical Issues and Citizen Rights in the Era of Digital Government Surveillance* (pp. 35–56). Hershey, PA: IGI Global. doi:10.4018/978-1-4666-9905-2.ch003

Ho, H. K. (2016). Embodying Difference on YouTube: Asian American Identity Work in "Shit Asian Dads Say". In A. Novak & I. El-Burki (Eds.), *Defining Identity and the Changing Scope of Culture in the Digital Age* (pp. 1–14). Hershey, PA: IGI Global. doi:10.4018/978-1-5225-0212-8.ch001

Hsu, C., & Wang, T. (2015). Composition of the Top Management Team and Information Security Breaches. In M. Cruz-Cunha & I. Portela (Eds.), *Handbook of Research on Digital Crime, Cyberspace Security, and Information Assurance* (pp. 116–134). Hershey, PA: IGI Global. doi:10.4018/978-1-4666-6324-4.ch008

Hussain, Z., & Pontes, H. M. (2018). Personality, Internet Addiction, and Other Technological Addictions: A Psychological Examination of Personality Traits and Technological Addictions. In B. Bozoglan (Ed.), *Psychological, Social, and Cultural Aspects of Internet Addiction* (pp. 45–71). Hershey, PA: IGI Global. doi:10.4018/978-1-5225-3477-8.ch003

Ingle, S., & Kuprevich, C. L. (2016). Workforce Development in Behavioral Healthcare and the Increased Use of Technology: Is It Working or Not? Are We Asking the Right Questions? In B. Baggio (Ed.), *Analyzing Digital Discourse and Human Behavior in Modern Virtual Environments* (pp. 40–59). Hershey, PA: IGI Global. doi:10.4018/978-1-4666-9899-4.ch003

Kang, Y., & Yang, K. C. (2016). Analyzing Multi-Modal Digital Discourses during MMORPG Gameplay through an Experiential Rhetorical Approach. In B. Baggio (Ed.), *Analyzing Digital Discourse and Human Behavior in Modern Virtual Environments* (pp. 220–243). Hershey, PA: IGI Global. doi:10.4018/978-1-4666-9899-4.ch012

Kidd, D., & Turner, A. J. (2016). The #GamerGate Files: Misogyny in the Media. In A. Novak & I. El-Burki (Eds.), *Defining Identity and the Changing Scope of Culture in the Digital Age* (pp. 117–139). Hershey, PA: IGI Global. doi:10.4018/978-1-5225-0212-8.ch008

Knibbs, C., Goss, S., & Anthony, K. (2017). Counsellors' Phenomenological Experiences of Working with Children or Young People who have been Cyberbullied: Using Thematic Analysis of Semi Structured Interviews. *International Journal of Technoethics, 8*(1), 68–86. doi:10.4018/IJT.2017010106

Koukovini, M. N., Papagiannakopoulou, E. I., Lioudakis, G. V., Dellas, N. L., Kaklamani, D. I., & Venieris, I. S. (2015). Privacy Compliance Requirements in Workflow Environments. In M. Cruz-Cunha & I. Portela (Eds.), *Handbook of Research on Digital Crime, Cyberspace Security, and Information Assurance* (pp. 157–177). Hershey, PA: IGI Global. doi:10.4018/978-1-4666-6324-4.ch011

Kovacevic, A., & Nikolic, D. (2015). Automatic Detection of Cyberbullying to Make Internet a Safer Environment. In M. Cruz-Cunha & I. Portela (Eds.), *Handbook of Research on Digital Crime, Cyberspace Security, and Information Assurance* (pp. 277–290). Hershey, PA: IGI Global. doi:10.4018/978-1-4666-6324-4.ch018

Kovacevic, A., & Nikolic, D. (2015). Cyber Attacks on Critical Infrastructure: Review and Challenges. In M. Cruz-Cunha & I. Portela (Eds.), *Handbook of Research on Digital Crime, Cyberspace Security, and Information Assurance* (pp. 1–18). Hershey, PA: IGI Global. doi:10.4018/978-1-4666-6324-4.ch001

Kowalsky, M. (2016). Analysis of Initial Involvement of Librarians in the Online Virtual World of Second Life. In B. Baggio (Ed.), *Analyzing Digital Discourse and Human Behavior in Modern Virtual Environments* (pp. 126–148). Hershey, PA: IGI Global. doi:10.4018/978-1-4666-9899-4.ch007

Kulesza, J. (2015). Surveillance, Privacy, and Due Diligence in Cybersecurity: An International Law Perspective. In M. Cruz-Cunha & I. Portela (Eds.), *Handbook of Research on Digital Crime, Cyberspace Security, and Information Assurance* (pp. 379–397). Hershey, PA: IGI Global. doi:10.4018/978-1-4666-6324-4.ch024

Lai, F. T., & Kwan, J. L. (2018). Socioeconomic Determinants of Internet Addiction in Adolescents: A Scoping Review. In B. Bozoglan (Ed.), *Psychological, Social, and Cultural Aspects of Internet Addiction* (pp. 127–145). Hershey, PA: IGI Global. doi:10.4018/978-1-5225-3477-8.ch007

Layton, R., & Watters, P. A. (2015). Indirect Attribution in Cyberspace. In M. Cruz-Cunha & I. Portela (Eds.), *Handbook of Research on Digital Crime, Cyberspace Security, and Information Assurance* (pp. 246–262). Hershey, PA: IGI Global. doi:10.4018/978-1-4666-6324-4.ch016

Lebraty, J., & Godé, C. (2015). Assessing the Performance of Decision Support Systems in Military Environment: The 3C Method. In A. Mesquita & C. Tsai (Eds.), *Human Behavior, Psychology, and Social Interaction in the Digital Era* (pp. 45–70). Hershey, PA: IGI Global. doi:10.4018/978-1-4666-8450-8.ch003

Lin, L., & Lipsmeyer, B. (2015). The Environmental and Technological Factors of Multitasking. In A. Mesquita & C. Tsai (Eds.), *Human Behavior, Psychology, and Social Interaction in the Digital Era* (pp. 1–20). Hershey, PA: IGI Global. doi:10.4018/978-1-4666-8450-8.ch001

Lin, T., Wu, Z., & Chen, Y. (2015). Using High-Frequency Interaction Events to Automatically Classify Cognitive Load. In A. Mesquita & C. Tsai (Eds.), *Human Behavior, Psychology, and Social Interaction in the Digital Era* (pp. 210–228). Hershey, PA: IGI Global. doi:10.4018/978-1-4666-8450-8.ch010

Ling, K. C., Ling, C. P., Zhimin, W., Hung, K. K., & Leong, L. H. (2017). The Impacts of Reactive Aggression and Friendship Quality on Cyberbullying Behaviour: An Advancement of Cyclic Process Model. *International Journal of Cyber Behavior, Psychology and Learning*, *7*(2), 49–71. doi:10.4018/IJCBPL.2017040105

Loutzenhiser, K. (2016). Public Administrators, School Safety, and Forms of Surveillance: Ethics and Social Justice in the Surveillance of Students' Disabilities. In R. Cropf & T. Bagwell (Eds.), *Ethical Issues and Citizen Rights in the Era of Digital Government Surveillance* (pp. 232–248). Hershey, PA: IGI Global. doi:10.4018/978-1-4666-9905-2.ch012

Macur, M., & Pontes, H. M. (2018). Individual Differences and the Development of Internet Addiction: A Nationally Representative Study. In B. Bozoglan (Ed.), *Psychological, Social, and Cultural Aspects of Internet Addiction* (pp. 221–235). Hershey, PA: IGI Global. doi:10.4018/978-1-5225-3477-8.ch012

Manthiou, A., Chiang, L. L., & Tang, L. R. (2015). Developing a Successful Facebook Fan Page Based on Costumers' Needs. In A. Mesquita & C. Tsai (Eds.), *Human Behavior, Psychology, and Social Interaction in the Digital Era* (pp. 189–209). Hershey, PA: IGI Global. doi:10.4018/978-1-4666-8450-8.ch009

Marín, J. M., Naranjo, J. Á., & Casado, L. G. (2015). Honeypots and Honeynets: Analysis and Case Study. In M. Cruz-Cunha & I. Portela (Eds.), *Handbook of Research on Digital Crime, Cyberspace Security, and Information Assurance* (pp. 452–482). Hershey, PA: IGI Global. doi:10.4018/978-1-4666-6324-4.ch029

Martinho, V. J. (2015). Analysis of the Cybercrime with Spatial Econometrics in the European Union Countries. In M. Cruz-Cunha & I. Portela (Eds.), *Handbook of Research on Digital Crime, Cyberspace Security, and Information Assurance* (pp. 483–499). Hershey, PA: IGI Global. doi:10.4018/978-1-4666-6324-4.ch030

Martins, M. J., Simão, A. M., Freire, I., Caetano, A. P., & Matos, A. (2016). Cyber-Victimization and Cyber-Aggression among Portuguese Adolescents: The Relation to Family Support and Family Rules. *International Journal of Cyber Behavior, Psychology and Learning*, *6*(3), 65–78. doi:10.4018/IJCBPL.2016070105

Masrom, M. (2016). E-Government, E-Surveillance, and Ethical Issues from Malaysian Perspective. In R. Cropf & T. Bagwell (Eds.), *Ethical Issues and Citizen Rights in the Era of Digital Government Surveillance* (pp. 249–263). Hershey, PA: IGI Global. doi:10.4018/978-1-4666-9905-2.ch013

McLaughlin, B., Hull, S., Namkoong, K., Shah, D., & Gustafson, D. H. (2016). We All Scream for Ice Cream: Positive Identity Negotiation in the Face of Cancer. In A. Novak & I. El-Burki (Eds.), *Defining Identity and the Changing Scope of Culture in the Digital Age* (pp. 81–98). Hershey, PA: IGI Global. doi:10.4018/978-1-5225-0212-8.ch006

McNeal, R. S., Schmeida, M., & Holmes, J. (2016). The E-Government Surveillance in the United States: Public Opinion on Government Wiretapping Powers. In R. Cropf & T. Bagwell (Eds.), *Ethical Issues and Citizen Rights in the Era of Digital Government Surveillance* (pp. 208–230). Hershey, PA: IGI Global. doi:10.4018/978-1-4666-9905-2.ch011

Millman, C. M., Winder, B., & Griffiths, M. D. (2017). UK-Based Police Officers' Perceptions of, and Role in Investigating, Cyber-Harassment as a Crime. *International Journal of Technoethics*, *8*(1), 87–102. doi:10.4018/IJT.2017010107

Mou, N. Z., & Islam, M. S. (2016). Re-Routing the Masculinity Myths in Bangladeshi Fashion Adverts: Identifying a New Wave among the Youths. In A. Novak & I. El-Burki (Eds.), *Defining Identity and the Changing Scope of Culture in the Digital Age* (pp. 170–191). Hershey, PA: IGI Global. doi:10.4018/978-1-5225-0212-8.ch011

Musa, W. (2015). Cybercrimes Technologies and Approaches. In M. Cruz-Cunha & I. Portela (Eds.), *Handbook of Research on Digital Crime, Cyberspace Security, and Information Assurance* (pp. 193–210). Hershey, PA: IGI Global. doi:10.4018/978-1-4666-6324-4.ch013

Myers, Z. R., Swearer, S. M., Martin, M. J., & Palacios, R. (2017). Cyberbullying and Traditional Bullying: The Experiences of Poly-Victimization Among Diverse Youth. *International Journal of Technoethics*, *8*(2), 42–60. doi:10.4018/IJT.2017070104

Nene, M. J., & Gupta, P. (2018). CyberPsycho Effect: A Critical Study on the Impact of Internet Addiction. In B. Bozoglan (Ed.), *Psychological, Social, and Cultural Aspects of Internet Addiction* (pp. 186–198). Hershey, PA: IGI Global. doi:10.4018/978-1-5225-3477-8.ch010

Niedt, G. (2016). Social Media Affordances and the Capital of Queer Self-Expression: Facebook, Ello, and the Nymwars. In A. Novak & I. El-Burki (Eds.), *Defining Identity and the Changing Scope of Culture in the Digital Age* (pp. 99–116). Hershey, PA: IGI Global. doi:10.4018/978-1-5225-0212-8.ch007

Novak, A. N., & Richmond, J. C. (2016). The Phrase Has Been Hijacked: Studying Generational Communication on Feminism through Social Media. In A. Novak & I. El-Burki (Eds.), *Defining Identity and the Changing Scope of Culture in the Digital Age* (pp. 156–169). Hershey, PA: IGI Global. doi:10.4018/978-1-5225-0212-8.ch010

Novoselova, V. (2016). Digitizing Consumer Activism: A Thematic Analysis of Jezebel.com. In A. Novak & I. El-Burki (Eds.), *Defining Identity and the Changing Scope of Culture in the Digital Age* (pp. 140–154). Hershey, PA: IGI Global. doi:10.4018/978-1-5225-0212-8.ch009

Oluwole, J. O., & Green, P. C. III. (2016). *Censorship and Student Communication in Online and Offline Settings* (pp. 1–622). Hershey, PA: IGI Global. doi:10.4018/978-1-4666-9519-1

Oravec, J. A. (2018). Cyber Bullying. In M. Khosrow-Pour, D.B.A. (Ed.), Encyclopedia of Information Science and Technology, Fourth Edition (pp. 1695-1703). Hershey, PA: IGI Global. doi:10.4018/978-1-5225-2255-3.ch148

Oriwoh, E., & Williams, G. (2015). Internet of Things: The Argument for Smart Forensics. In M. Cruz-Cunha & I. Portela (Eds.), *Handbook of Research on Digital Crime, Cyberspace Security, and Information Assurance* (pp. 407–423). Hershey, PA: IGI Global. doi:10.4018/978-1-4666-6324-4.ch026

Pereira, F. D., Veloso de Matos, M. A., & Sampaio, Á. M. (2015). Cyber-Crimes against Adolescents: Bridges between a Psychological and a Design Approach. In M. Cruz-Cunha & I. Portela (Eds.), *Handbook of Research on Digital Crime, Cyberspace Security, and Information Assurance* (pp. 211–230). Hershey, PA: IGI Global. doi:10.4018/978-1-4666-6324-4.ch014

Pina, A., Holland, J., & James, M. (2017). The Malevolent Side of Revenge Porn Proclivity: Dark Personality Traits and Sexist Ideology. *International Journal of Technoethics*, *8*(1), 30–43. doi:10.4018/IJT.2017010103

Raisinghani, M. S. (2015). Can Total Quality Management Exist in Cyber Security: Is It Present? Are We Safe? In M. Cruz-Cunha & I. Portela (Eds.), *Handbook of Research on Digital Crime, Cyberspace Security, and Information Assurance* (pp. 350–359). Hershey, PA: IGI Global. doi:10.4018/978-1-4666-6324-4.ch022

Reychav, I., & Sukenik, S. (2015). Cyberbullying: Keeping Our Children Safe in the 21st Century. In M. Cruz-Cunha & I. Portela (Eds.), *Handbook of Research on Digital Crime, Cyberspace Security, and Information Assurance* (pp. 77–98). Hershey, PA: IGI Global. doi:10.4018/978-1-4666-6324-4.ch006

Richet, J. (2015). How to Become a Cybercriminal?: An Explanation of Cybercrime Diffusion. In A. Mesquita & C. Tsai (Eds.), *Human Behavior, Psychology, and Social Interaction in the Digital Era* (pp. 229–240). Hershey, PA: IGI Global. doi:10.4018/978-1-4666-8450-8.ch011

Rive, P. B. (2016). Virtual Design Teams in Virtual Worlds: A Theoretical Framework using Second Life. In B. Baggio (Ed.), *Analyzing Digital Discourse and Human Behavior in Modern Virtual Environments* (pp. 60–86). Hershey, PA: IGI Global. doi:10.4018/978-1-4666-9899-4.ch004

Robson, G., & Olavarria, C. M. (2016). Big Collusion: Corporations, Consumers, and the Digital Surveillance State. In R. Cropf & T. Bagwell (Eds.), *Ethical Issues and Citizen Rights in the Era of Digital Government Surveillance* (pp. 127–144). Hershey, PA: IGI Global. doi:10.4018/978-1-4666-9905-2.ch007

Rodríguez-de-Dios, I., & Igartua, J. (2016). Skills of Digital Literacy to Address the Risks of Interactive Communication. *Journal of Information Technology Research*, *9*(1), 54–64. doi:10.4018/JITR.2016010104

Rosette, O. N., Kazemeyni, F., Aghili, S., Butakov, S., & Ruhl, R. (2016). Achieving Balance between Corporate Dataveillance and Employee Privacy Concerns. In R. Cropf & T. Bagwell (Eds.), *Ethical Issues and Citizen Rights in the Era of Digital Government Surveillance* (pp. 163–175). Hershey, PA: IGI Global. doi:10.4018/978-1-4666-9905-2.ch009

Rosewarne, L. (2017). "Nothing Crueler than High School Students": The Cyberbully in Film and Television. *International Journal of Technoethics*, *8*(1), 1–17. doi:10.4018/IJT.2017010101

Rowe, N. C. (2016). Privacy Concerns with Digital Forensics. In R. Cropf & T. Bagwell (Eds.), *Ethical Issues and Citizen Rights in the Era of Digital Government Surveillance* (pp. 145–162). Hershey, PA: IGI Global. doi:10.4018/978-1-4666-9905-2.ch008

Saini, A., Gaur, M. S., & Laxmi, V. (2015). A Taxonomy of Browser Attacks. In M. Cruz-Cunha & I. Portela (Eds.), *Handbook of Research on Digital Crime, Cyberspace Security, and Information Assurance* (pp. 291–313). Hershey, PA: IGI Global. doi:10.4018/978-1-4666-6324-4.ch019

Samli, R. (2018). A Review of Internet Addiction on the Basis of Different Countries (2007–2017). In B. Bozoglan (Ed.), *Psychological, Social, and Cultural Aspects of Internet Addiction* (pp. 200–220). Hershey, PA: IGI Global. doi:10.4018/978-1-5225-3477-8.ch011

Sarna, G., & Bhatia, M. (2016). An Approach to Distinguish Between the Severity of Bullying in Messages in Social Media. *International Journal of Rough Sets and Data Analysis, 3*(4), 1–20. doi:10.4018/IJRSDA.2016100101

Shahriar, H., & Devendran, V. (2015). Hijacking of Clicks: Attacks and Mitigation Techniques. In M. Cruz-Cunha & I. Portela (Eds.), *Handbook of Research on Digital Crime, Cyberspace Security, and Information Assurance* (pp. 135–145). Hershey, PA: IGI Global. doi:10.4018/978-1-4666-6324-4.ch009

Shang, G. N. (2016). The Self of the Camera: Popular Practices of Photography and Self-Presentation in the New Social Media. In A. Novak & I. El-Burki (Eds.), *Defining Identity and the Changing Scope of Culture in the Digital Age* (pp. 240–250). Hershey, PA: IGI Global. doi:10.4018/978-1-5225-0212-8.ch015

Shen, L. (2018). Treatment of Internet Addiction. In B. Bozoglan (Ed.), *Psychological, Social, and Cultural Aspects of Internet Addiction* (pp. 284–309). Hershey, PA: IGI Global. doi:10.4018/978-1-5225-3477-8.ch015

Smyth, S. J., Curran, K., & Mc Kelvey, N. (2018). Internet Addiction: A Modern Societal Problem. In B. Bozoglan (Ed.), *Psychological, Social, and Cultural Aspects of Internet Addiction* (pp. 20–43). Hershey, PA: IGI Global. doi:10.4018/978-1-5225-3477-8.ch002

Soares, E. C. (2015). The Gatekeepers of Cyberspace: Surveillance, Control, and Internet Regulation in Brazil. In M. Cruz-Cunha & I. Portela (Eds.), *Handbook of Research on Digital Crime, Cyberspace Security, and Information Assurance* (pp. 361–378). Hershey, PA: IGI Global. doi:10.4018/978-1-4666-6324-4.ch023

Sterner, G., & Felmlee, D. (2017). The Social Networks of Cyberbullying on Twitter. *International Journal of Technoethics*, *8*(2), 1–15. doi:10.4018/IJT.2017070101

Stodt, B., Wegmann, E., & Brand, M. (2016). Predicting Dysfunctional Internet Use: The Role of Age, Conscientiousness, and Internet Literacy in Internet Addiction and Cyberbullying. *International Journal of Cyber Behavior, Psychology and Learning*, *6*(4), 28–43. doi:10.4018/IJCBPL.2016100103

Sule, D. (2015). Forensic Readiness and eDiscovery. In M. Cruz-Cunha & I. Portela (Eds.), *Handbook of Research on Digital Crime, Cyberspace Security, and Information Assurance* (pp. 178–192). Hershey, PA: IGI Global. doi:10.4018/978-1-4666-6324-4.ch012

Sweeny, J. (2017). Gendered Violence and Victim-Blaming: The Law's Troubling Response to Cyber-Harassment and Revenge Pornography. *International Journal of Technoethics*, *8*(1), 18–29. doi:10.4018/IJT.2017010102

Switzer, J. S., & Switzer, R. V. (2016). Virtual Teams: Profiles of Successful Leaders. In B. Baggio (Ed.), *Analyzing Digital Discourse and Human Behavior in Modern Virtual Environments* (pp. 1–24). Hershey, PA: IGI Global. doi:10.4018/978-1-4666-9899-4.ch001

Tai, Z., & Liu, X. (2016). Virtual Ties, Perceptible Reciprocity, and Real-Life Gratifications in Online Community Networks: A Study of QQ User Groups in China. In B. Baggio (Ed.), *Analyzing Digital Discourse and Human Behavior in Modern Virtual Environments* (pp. 164–180). Hershey, PA: IGI Global. doi:10.4018/978-1-4666-9899-4.ch009

Tellería, A. S. (2016). Liquid Communication in Mobile Devices: Affordances and Risks. In B. Baggio (Ed.), *Analyzing Digital Discourse and Human Behavior in Modern Virtual Environments* (pp. 196–219). Hershey, PA: IGI Global. doi:10.4018/978-1-4666-9899-4.ch011

Valls-Prieto, J. (2015). Fighting Cybercrime and Protecting Privacy: DDoS, Spy Software, and Online Attacks. In M. Cruz-Cunha & I. Portela (Eds.), *Handbook of Research on Digital Crime, Cyberspace Security, and Information Assurance* (pp. 146–155). Hershey, PA: IGI Global. doi:10.4018/978-1-4666-6324-4.ch010

Walker, J. A., & Jeske, D. (2016). Understanding Bystanders' Willingness to Intervene in Traditional and Cyberbullying Scenarios. *International Journal of Cyber Behavior, Psychology and Learning*, *6*(2), 22–38. doi:10.4018/IJCBPL.2016040102

Wang, W., Cheung, B. C., Leung, Z. C., Chan, K., & See-To, E. W. (2017). A Social Media Mining and Analysis Approach for Supporting Cyber Youth Work. *International Journal of Knowledge and Systems Science*, *8*(2), 1–16. doi:10.4018/IJKSS.2017040101

Wigfall, J. (2016). Nothing Random about Taste: Toni Morrison and the Algorithmic Canon. In A. Novak & I. El-Burki (Eds.), *Defining Identity and the Changing Scope of Culture in the Digital Age* (pp. 43–55). Hershey, PA: IGI Global. doi:10.4018/978-1-5225-0212-8.ch004

Wright, M. F. (2015). Emerging Adults' Coping Strategies: Longitudinal Linkages to their Involvement in Cyber Aggression and Cyber Victimization. *International Journal of Cyber Behavior, Psychology and Learning*, *5*(2), 1–14. doi:10.4018/ijcbpl.2015040101

Wright, M. F. (2016). The Roles of Age, Gender, and Ethnicity in Cyberbullying. In A. Novak & I. El-Burki (Eds.), *Defining Identity and the Changing Scope of Culture in the Digital Age* (pp. 192–208). Hershey, PA: IGI Global. doi:10.4018/978-1-5225-0212-8.ch012

Wu, Z., Lin, T., Tang, N., & Wu, S. (2015). Effects of Display Characteristics on Presence and Emotional Responses of Game Players. In A. Mesquita & C. Tsai (Eds.), *Human Behavior, Psychology, and Social Interaction in the Digital Era* (pp. 130–145). Hershey, PA: IGI Global. doi:10.4018/978-1-4666-8450-8.ch006

Xie, T., Zhu, Y., Lin, T., & Chen, R. (2015). Modeling Human Behavior to Reduce Navigation Time of Menu Items: Menu Item Prediction Based on Markov Chain. In A. Mesquita & C. Tsai (Eds.), *Human Behavior, Psychology, and Social Interaction in the Digital Era* (pp. 162–187). Hershey, PA: IGI Global. doi:10.4018/978-1-4666-8450-8.ch008

Yang, C., Ha, L., Yun, G. W., & Chen, L. (2015). From Relationship to Information: A Study of Twitter and Facebook Usage in Terms of Social Network Size among College Students. In A. Mesquita & C. Tsai (Eds.), *Human Behavior, Psychology, and Social Interaction in the Digital Era* (pp. 241–258). Hershey, PA: IGI Global. doi:10.4018/978-1-4666-8450-8.ch012

Yartey, F. N., & Ha, L. (2015). Smartphones and Self-Broadcasting among College Students in an Age of Social Media. In A. Mesquita & C. Tsai (Eds.), *Human Behavior, Psychology, and Social Interaction in the Digital Era* (pp. 95–128). Hershey, PA: IGI Global. doi:10.4018/978-1-4666-8450-8.ch005

Yu, S. (2016). Using Mixed Methods to Understand the Positive Correlation between Fear of Cyberbullying and Online Interaction. *International Journal of Virtual Communities and Social Networking*, *8*(3), 29–36. doi:10.4018/IJVCSN.2016070103

Zhang, Y. (2016). The Personalized and Personal "Mass" Media – From "We-Broadcast" to "We-Chat": Reflection on the Case of Bi Fujian Incident. In A. Novak & I. El-Burki (Eds.), *Defining Identity and the Changing Scope of Culture in the Digital Age* (pp. 29–42). Hershey, PA: IGI Global. doi:10.4018/978-1-5225-0212-8.ch003

About the Authors

Ramona McNeal is an associate professor in the Department of Political Science at the University of Northern Iowa. Her chief research interests include e-government, telehealth, cybercrimes and campaign finance reform. She also studies the impact of technology on participation, including its relationship to voting, elections, and public opinion. She has published work in numerous journals including *Journal of Information Technology & Politics*, *Social Science Quarterly*, *Policy Studies Journal*, *Public Administration Review*, *Political Research Quarterly* and *State Politics and Policy Quarterly*. She is a co-author of *Digital Citizenship: The Internet, Society and Participation* (MIT Press, 2007) with Karen Mossberger and Caroline Tolbert.

Susan M. Kunkle is an assistant professor in the Department of Sociology at Kent State University where she teaches courses on crime, delinquency, and corrections. She earned a Doctor of Philosophy degree in political science, with an emphasis in justice studies and policy analysis from Kent State University. Dr. Kunkle spent over four decades in the justice system working in several juvenile court systems and with the federal bureau of prisons. Additionally, she has extensive experience in working with community based systems of care for offenders and children at-risk. She previously served as the chairperson and trustee of a regional system of detention and rehabilitation programs that provided services for six juvenile courts. She was a recipient of the Community Crime Prevention Award (Stark County Prosecutor's Office) and the Dawn Marie Hendershot Award for exemplary service on behalf of child victims (Stark County Victims' Rights Coalition). Dr. Kunkle's research interests include cyber aggression, correctional institutions, childhood victimization, and reentry courts. She has presented at numerous conferences including the American Society of Criminology and the Midwest Political Science Association.

Mary Schmeida is a public policy expert who has served in several key research positions in the public and private healthcare sectors. She received her Ph.D. in Political Science, Public Policy Program in 2005 from Kent State University, Ohio, USA; and holds a Master's of Science in Nursing degree in Psychiatric Mental-Health from Kent State, at which received a clinical grant award from the National Institute of Mental Health. She has held several faculty appointments in public and private Universities. Her chief research interests and publications are government reform; e-government surveillance; state and local cybercrime policies; campaign reform and voting behavior; climate change and environmental health; mental health policy, telehealth policy, Medicare and Medicaid policy. She has published work in a number of books and academic journals including *Government Information Quarterly*, *Administration and Policy in Mental Health* and *Mental Health Services Research*, and the *Journal of Health Care for the Poor and Underserved.*

Index

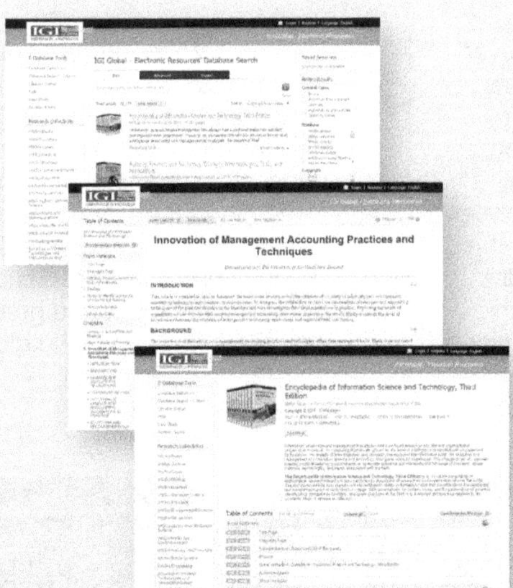

www.ingramcontent.com/pod-product-compliance
Lightning Source LLC
LaVergne TN
LVHW062125200926
848484LV00007B/1763